T0320701

Elementary Inelastic Radiation-Induced Processes

Library of Congress Cataloging-in-Publication Data

Elango, M. A. (Mart Aleksandrovich)
    [Elementarnye neuprugie radiatsionnye protsessy. English]
    Elementary inelastic radiation-induced processes / M. A. Elango;
translated by Albin Tybulewicz.
      p. cm.--(Translation series)
    Translation of: Elementarnye neuprugie radiatsionnye protsessy.
    Includes bibliographical references and index.
    ISBN 0-88318-799-X
    1. Condensed matter—Effect of radiation on. 2. Ionizing radiation.
    3. Inelastic cross sections. I. Title. II. Series:
Translation series (New York, NY)
QC173.4.C65E413   1990
530.4'16—dc20   90-48623
                                        CIP

Contents

Preface

Interaction with ionizing radiation is one of the most radical methods for altering the physical, chemical, and biological properties of condensed matter. This accounts for the continuing interest of researchers working in various branches of science in the mechanisms of the complex network of processes induced by the interaction of ionizing radiation with matter. The practical reasons for this interest are the need to identify the causes of undesirable changes in the properties of irradiated matter and to find ways for effective avoidance of these changes, on the one hand, and to isolate the desired changes in matter and find practical applications for such changes, on the other. This is interesting from the point of view of the theory of the condensed state because ionizing radiations provide means for creating and investigating states of matter far from thermodynamic equilibrium.

In most cases the macroscopically observable changes in the properties of the irradiated substance represent the results of complex chains of physical processes affecting both the electron and nuclear subsystems. The nature and effectiveness of these processes can be understood at the microscopic level if we separate the elementary steps in these processes so that they can be described in terms of quantum transitions. The method used to achieve this depends largely on the nature of the investigated radiation and matter, and on the purpose of the investigation. It is therefore frequently found that the separation into elementary stages varies so much from one concrete case to another that different cases are dealt with normally in different branches of science so that a clear relationship to one another and to the general hierarchy of the processes is lost.

This situation may be readily illustrated by the numerous monographs dealing with the subject. As a rule, the treatments dealing with the processes in the electron subsystem of matter (photoionization, ionization by electrons and other charged particles) ignore the processes occurring in the nuclear subsystem (transformation of the structure and creation of defects) and vice versa, in spite of the close relationship between these processes.

The purpose of the present monograph is to consider the most important elementary processes induced by ionizing radiations in condensed matter. We shall deal with all such processes and their inter-relationships and we shall provide an account not only of the fundamentals of the subject, but also of the latest progress. The ionizing radiations will be understood to be any radiations that alter the electronic state of a given object. They include not only high-energy nuclear radiations creating a wide range of various electron excitations and displacements of atoms, but also ultraviolet photons and in some cases even visible photons capable of effective creation of electron excitations in an object.

Attention is concentrated on the simplest quantitative description of the causes and effectiveness of the processes induced by such radiations in atoms, molecules, and particularly in solids, together with elementary derivations of the most important relationships. Descriptions are also given of the investigation methods based on inelastic interactions of radiations with matter and also of the principal radiation-induced changes in the physical, chemical, and biological properties of condensed matter, with an indication of the causal relationships between these changes and the elementary processes initiated by ionizing radiations. The treatment is based on fundamental concepts from quantum mechanics, statistical physics, and solid-state physics.

The author hopes that this type of presentation will be useful above all to experimentalists (physicists, chemists, biologists) working on changes in the properties of condensed matter resulting from the interaction with various ionizing radiations, and also to senior undergraduates and postgraduates in the relevant specialties.

The author is deeply grateful to Professor V. V. Khizhnyakov and to Dr. A. A. Elango for reading the manuscript and helpful comments. The author is indebted to Professor V. I. Ivanov for a critical reading of the manuscript and to Professor V. S. Vavilov for encouraging the writing of this book.

Chapter 1

Selected general concepts

1.1. Elementary process and its probability

Ionizing radiations interacting with condensed matter alter its structure and various properties. The changes are the result of a redistribution of electrons and atomic nuclei in matter considered in terms of energy and spatial coordinates. Such a redistribution begins with a primary inelastic interaction of a particle or a photon penetrating a body with electrons or nuclei in that body, which results in a transfer of an energy at some point in the body. The primary excitation created in this way can be described either as a pair consisting of a quasifree electron and a hole in some electron shell of some atom or as a pair consisting of a fast atom or an ion knocked out from its initial position in a body and an unoccupied site (vacancy).

The subsequent interaction with the environment degrades such excitation and secondary processes involve division of the energy into smaller portions and creation of a number of excitations with lower energies which then spread out over a certain volume. The final result is that the bulk of the transmitted energy becomes distributed more or less uniformly over the whole body and this increases somewhat its temperature. A small fraction of such secondary excitations is, however, stable or metastable (localized electrons and holes, interstitial atoms and vacancies, free radicals) and it is manifested by changes in the various macroscopic properties of the irradiated body which are retained for a long time (the meaning of "long time" must naturally be specified in each case) after irradiation.

The present book will attempt to provide a unified analysis of the mechanisms and consequences of primary and secondary processes participating in the chain of transitions described above. It is assumed that each complex process can be represented by a sum of "elementary" processes. It is postulated that a given physical object (atom, molecule, or condensed matter) has many states and that an elementary process is a transition (between two states) which we are unable (or deem unnecessary) to separate into its component parts. The evolution of an object with time is described by the basic kinetic equation

$$\frac{dP_n(t)}{dt} = \sum_m \left[W_{nm}P_m(t) - W_{mn}P_n(t) \right], \qquad (1.1)$$

where $P_n(t)$ is the probability that at a moment t an object is in a state n, whereas m is a state which an object can reach by transition from the state n. The coefficients of proportionality W_{mn} are called the probabilities of $n \rightarrow m$ transitions and are therefore the probabilities of elementary processes. They determine the reciprocal lifetime τ_{mn}^{-1} of an object which is in a state n before its transition to a state m:

$$W_{mn} = \tau_{mn}^{-1}. \tag{1.2}$$

If this transition is spontaneous, the probability W_{mn} describes fully its efficiency. In the case of induced transitions when the efficiency of the process depends also on the flux of the incident (inducing) agents

$$S = Nv, \tag{1.3}$$

where N is the concentration of such agents and v is their velocity, it is frequently more convenient to replace the probability with the cross section σ of the process related to the probability by

$$\sigma_{mn} = W_{mn}/S = (B/Av\tau_{mn}N)^{-1}. \tag{1.4}$$

The first methodological problem which has to be solved before we consider the processes of interest to us is the selection between the quantum and classical descriptions of the interaction. In the classical description the path of a particle as well as its energy and momentum are defined exactly. In the quantum case the particles have wave properties, so that their paths and momenta cannot be defined exactly simultaneously, and the phenomena of diffraction and interference play an important role in the interaction of particles.

The simplicity and clarity of the classical description is the reason for its preference over the quantum description in all those cases when the precision of the results obtained with its aid is sufficient. Two conditions can be used to identify those cases which can be described classically: (1) the uncertainty of the coordinates of a particle, related to its wavelength λ, should be less than the size of the interaction region a, i.e.,

$$\lambda = \hbar/p \ll a, \tag{1.5}$$

where p is the momentum of the particle, $\hbar = h/2\pi$, and h is the Planck constant; (2) the uncertainty of the scattering angle $\delta\theta$ of the incident particle, due to diffraction of its wave by an object, should be small compared with the value of θ itself, i.e.,

$$\delta\theta \approx \lambda/a \ll \theta. \tag{1.6}$$

If we bear in mind that $p = (2M_1E_0)^{1/2}$, where $E_0 = M_1v_0^2/2$ is the kinetic energy of the incident particle, and M_1 and v_0 are the mass and velocity of this particle, and if we introduce a characteristic energy

$$E_1' = \frac{\hbar^2}{2M_1a^2},$$

then instead of Eq. (1.5) we obtain the condition

$$E_0 \gg E_1', \tag{1.7}$$

whereas (1.6) is replaced with

$$\theta^2 E_0 \gg E_1'. \tag{1.8}$$

TABLE 1.1. Values of E_1' for some collisions.[51]

Incident particle	Target atom	E_1', eV	Incident particle	Target atom	E_1', eV
Electron	any	10	ion ($Z_1 = 1$, $M_1 = 100$)	Li	10^{-4}
Proton	Li	10^{-2}		U	10^{-5}
	U	10^{-3}			

Clearly, the condition (1.6) is generally more stringent than the condition (1.5).

The values of E_1' are given in Table 1.1 for several types of collisions. Although in the case of collisions of particles with atomic nuclei the quantity E_1' is small, we can always find a sufficiently small angle θ for which the condition (1.8) is not satisfied. However, the collisions of interest to us involve nuclei and these collisions, for which the transferred energy amounts to 1 eV or more (see Sec. 3.1), can quite justifiably be regarded as classical. It also follows from Table 1.1 that the scattering of low-energy electrons by other electrons requires a quantum-mechanical analysis. By definition, this is required also in the case of inelastic processes involving photons.

Let us consider the classical pattern of the scattering of an agent B, which has a mass M_1 and a velocity v_0, by an object A which can be regarded as an immobile center of a force field $U(r)$ (Fig. 1.1). We shall define the impact parameter b as the distance between A and the initial line of motion of the agent B, i.e., a distance at which B would have passed A in the absence of the force field. The interaction deflects B from its initial direction by an angle θ. In view of the spherical symmetry of the field $U(r)$, the orbit of B lies in one plane and is symmetric relative to a straight line AB_1 drawn from A to the nearest point on the orbit. The instantaneous position of B is described by a polar angle φ and a radius r. We shall write down (allowing for the curvilinear nature of the path of B) laws of conservation of the angular momentum

$$M_1 r^2 \frac{d\varphi}{dt} = M_1 v_0 b \tag{1.9}$$

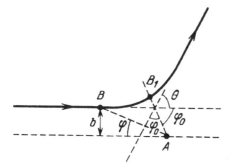

FIG. 1.1. Path of a particle B scattered by an immobile force-field center A.

and of the energy

$$\frac{M_1}{2}\left[\left(\frac{dr}{dt}\right)^2 + r^2\left(\frac{d\varphi}{dt}\right)^2\right] + U(r) = \frac{M_1 v_0^2}{2}. \tag{1.10}$$

The substitution of Eq. (1.9) into Eq. (1.10) gives

$$\frac{dr}{dt} = \pm\left(v_0^2 - \frac{b^2 v_0^2}{r^2} - \frac{2U(r)}{M_1}\right)^{1/2}. \tag{1.11}$$

If we now divide Eq. (1.9) by Eq. (1.11), we obtain

$$\frac{d\varphi}{dr} = \pm v_0 b \left/ \left[r^2\left(v_0^2 - \frac{b^2 v_0^2}{r^2} - \frac{2U(r)}{M_1}\right)^{1/2}\right]\right. . \tag{1.12}$$

We shall use φ_0 for the angles at which the asymptotes of the orbit intersect the straight line AB_1. It then follows from Fig. 1.1 and Eq. (1.12) that

$$|\theta = |\pi - 2\varphi_0| = \pi - 2v_0 b \int_{r_{min}}^{\infty} dr \left/ \left[r^2\left(v_0^2 - \frac{b^2 v_0^2}{r^2} - \frac{2U(r)}{M_1}\right)^{1/2}\right]\right., \tag{1.13}$$

where r_{min} is the distance of closest approach between A and B, found from Eq. (11) by substituting $dr/dt = 0$.

The expression (1.13) defines the function $b(\theta)$ which can be used to determine the cross section for the scattering into an angular element $d\theta$:

$$d\sigma = 2\pi b \, db = 2\pi b(\theta)\left|\frac{db(\theta)}{d\theta}\right| d\theta = \frac{b(\theta)}{\sin\theta}\left|\frac{db(\theta)}{d\theta}\right| d\Omega, \tag{1.14}$$

where $d\Omega = 2\pi \sin\theta \, d\theta$ is an element of the solid angle.

In the frequently encountered case of an electrostatic interaction between A and B, i.e., in the Coulomb scattering case, we have

$$U(r) = \frac{Z_A Z_B e^2}{r}, \tag{1.15}$$

where e is the electron charge, whereas Z_A and Z_B are the charges of A and B in units of e.

The substitution of Eq. (1.15) in Eq. (1.13) gives, after some transformations, the expression

$$b = \frac{Z_A Z_B e^2}{M_1 v_0^2 \sin(\theta/2)}, \tag{1.16}$$

which can be substituted in Eq. (1.14) to obtain the Rutherford formula

$$\frac{d\sigma}{d\Omega} = \left(\frac{Z_A Z_B e^2}{2M_1 v_0^2}\right)^2 \sin^{-4}\frac{\theta}{2} = \frac{a_C^2}{4}\sin^{-4}\frac{\theta}{2}, \tag{1.17}$$

where $a_C = Z_A Z_B e^2/(M_1 v_0^2) = b\sin(\theta/2)$ is the effective Coulomb scattering radius.

It should be pointed out that in the Coulomb scattering case the condition (1.7) can be rewritten (allowing for $a = a_C$) in the form

$$\frac{4Z_A Z_B e^2}{\hbar v_0} \gg 1. \tag{1.18}$$

If the object A is immobile (as is frequently found) only until the interaction of B, then Eqs. (1.13) and (1.17) give the cross section as a function of the angle of scattering in a system of center-of-inertia coordinates. The cross section can be calculated by adopting a laboratory coordinate system. However, in the important case when the mass of A is large compared with M_1 (i.e., when A is "almost immobile"), the above expressions retain their original form in the laboratory system if we replace M_1 by the reduced mass $\mu = M_1 M/(M_1 + M)$, where M is the mass of A.

In the quantum approach the basic kinetic equation (1.1) is valid if n denotes a separate quantum state of the system and the coefficients c_n in the expansion of the wave function of a completely isolated system

$$F = \sum_n c_n \Phi_n \tag{1.19}$$

have random phases at any moment in time. Here, $\{\Phi_n\}$ is a complete orthonormalized system of stationary wave functions, satisfying the Schrödinger equation

$$H\Phi_n = E_n \Phi_n, \tag{1.20}$$

where H is the Hamiltonian of the system and E_n is the energy of the state n.

In this approach the concept of an elementary process acquires a particularly clear meaning: it is a quantum transition of an object A from a state n (which we shall regard as initial and describe by a wave function Φ_0) to a state m (which we shall regard as final and describe by a wave function Φ_f) under the influence of an external perturbation by an agent B, described by the interaction Hamiltonian H_{AB}. According to perturbation theory, the probability of such a transition per unit time is described by the Fermi golden rule

$$W_{f0} = \frac{2\pi}{\hbar} |M_{f0}|^2 \delta(E_f - E_0), \tag{1.21}$$

where E_f and E_0 are the energies of the system $A + B$ before and after the interaction; δ is the Dirac delta function; M_{f0} represents the matrix element of a transition between the states f and 0:

$$M_{f0} = \int \Phi_f^* H_{AB} \Phi_0 \, d\tau \equiv \langle f | H_{AB} | 0 \rangle,$$

where Φ^* is the complex conjugate of Φ; $d\tau$ is an element of the phase volume.

Since the wave functions Φ_f and Φ_0 are orthonormalized ($\int \Phi_f \Phi_0 \, d\tau = \delta_{f0}$), the quantity M_{f0} is a measure of the degree of preparation for a transition from the initial state of A to its final state under the influence of an agent B.

The delta function in Eq. (1.20) ensures energy conservation in the process under consideration. If the final and/or initial state belongs to a continuous group of states, a transition may involve all the states characterized by $E_f = E_0$. If the matrix elements of all such transitions are approximately the same, the Fermi golden rule can be rewritten in the form

$$dW_{f0} = \frac{2\pi}{\hbar} |M_{f0}|^2 dp(E_f),$$ (1.22)

where $dp(E_f)$ is an element of the density of the final states (i.e., of the number of the states per unit energy interval) at which transitions can terminate.

In this monograph we shall frequently consider the cases when the final state is a free particle. The number of states of such particles with the momentum between p and $p + dp$ within a solid angle element $d\Omega$ is

$$dN = \frac{Vp^2 \, dp \, d\Omega}{(2\pi\hbar)^3},$$

where V is the spatial volume. Hence, the density-of-states function becomes

$$dp(E) = \frac{dN}{dE} = \frac{Vp^2}{(2\pi\hbar)^3} \frac{dp}{dE} d\Omega.$$ (1.23)

The density of states of particles is largely determined by their reciprocal dispersion $(dE/dp)^{-1}$.

1.2. Electron and nuclear subsystems

In the coordinate representations the wave functions Φ_n are, in principle, functions of the positions r_i of all the microparticles in a given system. If the system being ionized is more complex than the hydrogen atom, the Schrödinger equation (1.20) cannot be solved exactly and the exact wave functions cannot be obtained. The problem is simplified and clear results are obtained if Φ_n is simplified right from the beginning. The main tendency is then to factorize Φ_n so that only the minimum number of factors (in the limit, one factor) contains coordinates sensitive to the investigated perturbations of the object. If, for example, a perturbation alters significantly only the coordinates $r_1, r_2, ..., r_k$, whereas the influence on the coordinates $r_{k+1}, ..., r_n$ during the time of such a change can be ignored, it is expedient to represent Φ_n in the form

$$\Phi_n(r_1, r_2, ..., r_n) = \Phi_{n1}(r_1, ..., r_k) \Phi_{n2}(r_{k+1}, ..., r_n).$$ (1.24)

Then, the perturbation operator influences only Φ_{n1} and the matrix element (1.21) can be transformed as follows:

$$M_{f0} = \int \Phi_{1f}^* \Phi_{2f}^* H_{AB} \Phi_{10} \Phi_{20} \, d\tau_1 \, d\tau_2$$

$$= \int \Phi_{1f}^* H_{AB} \Phi_{10} \, d\tau_1 \cdot \int \Phi_{2f}^* \Phi_{20} \, d\tau_2$$

$$= M_{1f0} \int \Phi_{2f}^* \Phi_{20} \, d\tau_2$$

$$= M_{1f0} F_{2f0},$$ (1.25)

where $d\tau_1 = dr_1, ..., dr_k$, $d\tau_2 = dr_{k+1}, ..., dr_n$.

In this case the system $r_1,...,r_k$ is called a fast subsystem, whereas the system $r_{k+1},...,r_n$ is a slow subsystem. The energy of the fast subsystem plays the role of the potential energy in the effective Hamiltonian of the slow subsystem, so that a transition in the former always alters the Hamiltonian of the latter. Frequently however, these changes are slight. We then have $\Phi_{2f} \approx \Phi_{20}$, $F_{2f0} \approx \delta_{f0}$, and

$$M_{f0} \approx M_{1f0}. \qquad (1.26)$$

If we cannot ignore the change in the Hamiltonian of the slow subsystem in the course of a transition in the fast subsystem, then Φ_{2f} and Φ_{20} belong to different sets of orthonormalized wave functions and $F_{2f0} \neq \delta_{f0}$. In this case one speaks of a sudden perturbation of the slow subsystem. Such a perturbation gives rise to characteristic excitations of the slow subsystem. This process is frequently called the shake-off of these excitations.

The first concrete step in the simplification procedure described above is usually separation of the motion of electrons and nuclei in matter. This is justified by the large difference between the masses of electrons m and nuclei M ($M/m \geqslant 1840$). If an atom is ionized, this ratio allows us to ignore the recoil energy of a nucleus when an electron is removed from an atom. In discussing molecules and solids, when the distribution of nuclei relative to one another plays an important role, the above relationship makes it possible to use the Born–Oppenheimer or adiabatic approximation:

$$\Phi_n(r,R) = \Psi_n(r;R)\varphi_n(R), \qquad (1.27)$$

where r and R are the coordinates of electrons and nuclei of the system, respectively. The function $\Psi_n(r;R)$ is the wave function of the electron subsystem: it depends only parametrically on nuclear coordinates. The function $\varphi_n(R)$ is the wave function of the nuclear subsystem in the average electron field, i.e., it describes vibrations of the nuclei.

When the processes in the electron subsystem are considered, this approximation allows us to ignore the influence of H_{AB} on $\varphi_n(R)$. Then, the substitutions $\Phi_1 \to \Psi$ and $\Phi_2 \to \varphi$ transform Eq. (1.25) into

$$M_{f0} = \int M^e_{f0}(R)\varphi^*_f(R)\varphi_0(R)dR,$$

where

$$M^e_{f0} = \int \Psi^*_f H_{AB}\Psi_0 \, dr.$$

Let us assume that in the first approximation the quantity $M^e_{f0}(R)$ is independent of R near the equilibrium value of the latter R_0 [or, in other words, we can expand $M^e_{f0}(R)$ as a series near the point R_0 and retain only the first term of the series]. This gives

$$M_{f0} = M^e_{f0}(R_0) \int \varphi^*_f(R)\varphi_0(R)dR. \qquad (1.28)$$

This is known as the Condon approximation. In this approximation the probability of an electron transition is found by multiplying the electron matrix element and the overlap integral of the vibrational wave functions. The square of this integral is known as the Franck–Condon factor.

Since $\varphi_n(R)$ is the wave function of the nuclei in the average field created by the electrons and this field generally changes as a result of an electron transition, φ_f and φ_0 are the solutions of different Schrödinger equations and are not orthonormalized. The difference between them is particularly important in the case of small molecules. However, sometimes a change in the state of one electron in a many-electron system may have practically no influence on the potential describing the motion of the nuclei. We then have $\int \varphi_f^* \varphi_0 \, dR = \delta_{f0}$ and we find that the transitions involving a change in the vibrational state of the system are forbidden in the Condon approximation, whereas the probabilities of the allowed transitions are governed by the electron matrix element.

The physical meaning of the Condon approximation can be described as follows: a change in the states of nuclei in the course of an electron transition is unlikely, but the nuclear subsystem is perturbed suddenly and it emits (shakes off) elementary excitations of the nuclear subsystem in the form of its vibrations.

1.3. Derivatives of the basic kinetic equation

The basic kinetic equation (1.1) also has to be simplified in practice when describing complex systems. In the majority of cases of interest to us the quantity n can have a large number of values with a continuous or a quasicontinuous distribution in terms of one or more parameters, the set of which we will denote by s. Then, a many-particle relaxing system can be described usefully by a distribution function $f(t,s)$ expressing the probability that a state with the given value of s is occupied at a moment t. We shall normalize this function so that

$$\int f(t,s)\,ds = N(t), \tag{1.29}$$

where $N(t)$ is the average concentration of particles in the space of the quantity s, i.e., $N(t)\,ds$ is the average number of particles in a unit volume ds of this space.

We shall rewrite Eq. (1.1) for this function in the integral form:

$$\frac{df(t,s)}{dt} = \int_{\Delta s} \left[W(s + \Delta s, \Delta s)\, f(t, s + \Delta s) - W(s, \Delta s)\, f(t,s) \right] d\Delta s. \tag{1.30}$$

Here, $W(s, \Delta s)$ is the probability of a change in a parameter s in an elementary relaxation event by an amount Δs. The derivative df/dt can be rewritten as follows:

$$\frac{df(s,t)}{dt} = \frac{\partial f(t,s)}{\partial t} + \frac{\partial f(t,s)}{\partial s}\frac{ds}{dt} = \frac{\partial f(t,s)}{\partial t} + v_s \frac{\partial f(t,s)}{\partial s}, \tag{1.31}$$

where v_s is the velocity of particle drift along the coordinate s.

We shall simplify Eq. (1.30) by dividing the set of Δs into two parts. In the first we shall include small values of Δs, which are much less than the whole range of s in the course of relaxation ($\Delta s \ll s$). This corresponds most closely to the usual understanding of the process as a sequence of many small changes in the state of the system. The time $\tau_r = [\int_{\Delta s \ll s} W(s, \Delta s)\, d\Delta s]^{-1}$ averaged over all such processes is called the relaxation time of the particles. In the second part of the set we retain the large values of Δs ($\Delta s \approx s$). The corresponding transitions can be regarded as events of creation and loss of particles characterized by a coordinate s. The time $\tau_s = [\int_{\Delta s \approx s} W(s, \Delta s)\, d\Delta s]^{-1}$ averaged over all the loss processes will be called the lifetime of the particles. We shall define the particle generation $G(t,s)$ and loss $T(t,s)$ rates as follows:

$$G(t,s) = \int_{\Delta s \approx s} W(s + \Delta s, \Delta s)\, f(t, s + \Delta s)\, d\Delta s, \tag{1.32a}$$

$$T(t,s) = f(t,s) \int_{\Delta s \approx s} W(s, \Delta s)\, d\Delta s = \frac{f(t,s)}{\tau_s}. \tag{1.32b}$$

In the case when $\Delta s \ll s$, the first term in the integral in Eq. (1.30) can be expanded as a series:

$$W(s + \Delta s, \Delta s)\, f(t, s + \Delta s) = W(s, \Delta s)\, f(t,s) + \Delta s \frac{\partial}{\partial s}\, [\, W(s, \Delta s)\, f(t,s)\,]$$

$$+ \frac{1}{2}\, (\Delta s)^2 \frac{\partial^2}{\partial s^2}\, [\, W(s, \Delta s)\, f(t,s)\,] + \cdots .$$

We then have

$$\int_{\Delta s \ll s} \left[\Delta s \frac{\partial}{\partial s}\, (Wf) + \frac{1}{2}\, (\Delta s)^2 \frac{\partial^2}{\partial s^2}\, (Wf) \right] d\Delta s$$

$$= \frac{\partial}{\partial s} \left\{ f \int \Delta s W d\Delta s + \frac{1}{2} \frac{\partial}{\partial s} \left[f \int (\Delta s)^2 W d\Delta s \right] \right\}, \tag{1.33}$$

where $W \equiv W(s, \Delta s)$ and $f \equiv f(t, s)$.

Using Eqs. (1.32) and (1.33), we find instead of Eq. (1.30),

$$\frac{df}{dt} = G - T + \frac{\partial}{\partial s} \left[Af + \frac{\partial}{\partial s}\, (Bf) \right] = G - T + \operatorname{div} j_s, \tag{1.34}$$

where

$$A = \int_{\Delta s \ll s} \Delta s W d\Delta s, \tag{1.35a}$$

$$B = \frac{1}{2} \int_{\Delta s \ll s} (\Delta s)^2 W d\Delta s, \tag{1.35b}$$

$$j_s = Af + \frac{\partial}{\partial s}\, (Bf), \tag{1.35c}$$

where j_s is the flux of particles along the coordinate s. If the relaxation process is random, then W is an isotropic function, $\Delta s W$ is an even function, and $A = 0$.

Equation (1.34) is known as the Fokker–Planck equation. It can be used to describe the process of relaxation in terms of the average characteristics of its elementary stages.

The Fokker–Planck equation can be simplified even further if A and B are independent of s or if we can use the values $A = \overline{\Delta s}/\tau_r = v_s$ and $B = (\frac{1}{2}) \times (\overline{\Delta s})^2/\tau_r = D_s$ averaged over s. (Here, D_s is the diffusion coefficient of particles along the coordinate s.) It then follows from Eqs. (1.34) and (1.31) that

$$\frac{\partial f}{\partial t} = D_s \frac{\partial^2 f}{\partial s^2} - v_s \frac{\partial f}{\partial s} + G - T. \tag{1.36}$$

We have thus derived a diffusion (or a diffusion–drift) equation. A comparison of Eqs. (1.34) and (1.36) demonstrates that the Fokker–Planck equation is essentially a diffusion equation with variable coefficients and the diffusion equation is a Fokker–Planck equation with constant coefficients.

Let us consider the general case of constant number of particles ($G = T = 0$) represented by independent parameters in the form of the coordinate r and the velocity $\mathbf{v}$ (or the momentum $\mathbf{p}$). It follows from Eqs. (1.30) and (1.31) that the distribution function is described by

$$\frac{\partial f}{\partial t} + \mathbf{v}\frac{\partial f}{\partial r} + \frac{F}{M}\frac{\partial f}{\partial \mathbf{v}} = St, \tag{1.37}$$

where

$$St = \int [W(\mathbf{v},r|\mathbf{v}',r')\,f(\mathbf{v},r,t) - W(\mathbf{v}',r'|\mathbf{v},r)\,f(\mathbf{v}',r',t)]d\mathbf{v}'dr'.$$

Here, St is known as the collision integral; $W(\mathbf{v},r|\mathbf{v}',r')$ is the probability of an elementary event of collision of a particle traveling at a velocity $\mathbf{v}$ at a point r, which alters these quantities to $\mathbf{v}'$ and r'; $F = M\,d\mathbf{v}/dt$ is the force exerted by an external field; M is the mass of a particle.

Equation (1.37) is known at the Boltzmann equation. It is frequently used to describe the behavior of the band electrons and holes in solids, and also of electrons and ions in a plasma. It is more informative than the Fokker–Planck equation because it does not postulate either a relatively small change in the parameters of the particles nor their averaging over many elementary scattering events. However, the general Boltzmann equation is usually far too complex for practical applications. In most cases this equation is employed after two additional simplifications. First, it is assumed that the principle of detailed equilibrium applies, i.e., that the probabilities of forward and reverse processes are equal: $W(\mathbf{v},r|\mathbf{v}',r') = W(\mathbf{v}',r'|\mathbf{v},r)$. Secondly, a relaxation time $\tau_r = [\int W(\mathbf{v},r|\mathbf{v}',r')d\mathbf{v}'\,dr']^{-1}$ is introduced. Then, the collision integral St becomes

$$St = -\frac{f - f_0}{\tau_r}, \tag{1.38}$$

where f_0 is the equilibrium distribution function.

If f in Eq. (1.38) is expanded in terms of small changes in $\mathbf{v}$ and $\mathbf{r}$, we again obtain the diffusion equation.

Chapter 2

Electron subsystem

2.1. Photoionization of atoms

The photoionization of atoms will be understood here to be a process in which the incident particle is a photon absorbed by an object:

$$A + \hbar\omega \rightarrow A^{n+} + ne. \tag{2.1}$$

Here, ω is the angular frequency of the incident photon ($\omega = 2\pi\nu$, $\lambda\nu = c$, ν is the photon frequency, λ is the photon wavelength, and c is the velocity of light); n is the degree of ionization.

The simplest variant of the process described by Eq. (2.1) is single ionization ($n = 1$):

$$A + \hbar\omega \rightarrow A^+ + e. \tag{2.2}$$

In the case of many-electron atoms and more complex physical systems, the exact electron wave function Ψ is far too complex even when it can be separated from the wave function of the whole system (see Sec. 1.2). We shall simplify it further by adopting the concept of fast and slow subsystems bearing in mind that in the majority of cases the perturbation of the electron subsystem in matter alters considerably the motion of a small number of electrons (in the limit it changes the motion of one electron). We can then assume that the motion of an electron undergoing a transition called a "transition electron" (fast subsystem) is independent of the motion of other electrons (slow subsystem) and the influence of the other electrons can be described by some average field. In this approximation the wave function of the electron system can be written in the form

$$\Psi(r) = A\psi(r_1)\Psi'(r_2, r_3, ..., r_k), \tag{2.3}$$

where r_1 are the coordinates of the transition electron; $r_2, ..., r_k$ are the coordinates of the remaining electrons; Ψ' is the wave function of the system without the transition electron; A is an antisymmetrizing operator ensuring constancy of $\Psi(r)$ under transposition of electrons $r_i \rightleftarrows r_j$, i.e., $A^2 = 1$. Then, the electron matrix element is described by

11

$$M_{f0}^e = \int \Psi_f'(r) H_{AB} \Psi_0(r) dr$$

$$= \int \psi_f^*(r_1) H_{AB} \psi_0(r_1) dr_1 \int \Psi_f'^* \Psi_0' \, dr'$$

$$= M_{f0}^{0e} \int \Psi_f'^* \Psi_0' dr', \tag{2.4}$$

where M_{f0}^{0e} is the one-electron matrix element of the transition. The functions Ψ_f' and Ψ_0' are, generally speaking, solutions of different Schrödinger equations (if in the case of single ionization, Ψ_0' represents a system of k electrons, then Ψ_f' applies to a system of $k - 1$ electrons) and are not orthonormalized. However, the difference between the corresponding potentials is frequently negligible. We can then assume that $\int \Psi_f'^* \Psi_0' \, dr' = \delta_{f0}$ and M_{f0}^e is equal to M_{f0}^{0e}. This approximation is called the one-electron approach. It follows from Eqs. (1.22), (1.28), and (2.4) that the golden rule for this approximation is

$$dW_{f0} = \frac{2\pi}{\hbar} \left| \int \psi_f^* H_{AB} \psi_0 \, dr \right|^2 \left| \int \Psi_f'^* \Psi_0' dr' \right|^2 \left| \int \varphi_f^* \varphi_0 dR \right|^2 d\rho(E_f)$$

$$\approx \frac{2\pi}{\hbar} \left| \int \psi_f^* H_{AB} \psi_0 dr \right|^2 d\rho(E_f). \tag{2.5}$$

The one-electron wave functions ψ of a system of electrons determine, via the Schrödinger equation (1.20), the stationary states of the system corresponding to specific energy levels. In the hydrogenlike approximation a level with the principal quantum number n has the energy

$$E_n = -\frac{me^4}{\hbar^2} \frac{Z^2}{2n^2} = -E_B \frac{Z^2}{2n^2}, \tag{2.6}$$

the electron velocity is

$$v_n = \frac{e^2}{\hbar} \frac{Z}{n} = v_B \frac{Z}{n}, \tag{2.7}$$

the average distance of an electron from the center of the force field is

$$a_n = \frac{\hbar^2}{me^2} \frac{n}{Z} = a_B \frac{n}{Z}, \tag{2.8}$$

and the time constant is

$$\tau_n = \frac{a_n}{v_n} = \frac{\hbar^3}{me^4} \frac{n^2}{Z^2} = \tau_B \frac{n^2}{Z^2}, \tag{2.9}$$

where Z is the charge of the nucleus of an atom in units of e; m is the rest mass of an electron; $E_B = me^2/\hbar^2 \approx 27.2$ eV, $v_B = e^2/\hbar \approx 2.19 \times 10^8$ cm/s, $a_B = \hbar^2/me^2 \approx 5.29 \times 10^{-9}$ cm, $\tau_B = \hbar^3/me^4 \approx 2.42 \times 10^{-17}$ s are the atomic (Bohr) units of the corresponding quantities. We note that $v_B = \alpha c$, where $\alpha = e^2/\hbar c \approx \frac{1}{137}$ is the fine-

TABLE 2.1. Energy of one-electron ionization of electron shells of inert-gas atoms (eV), taken from Ref. 125.

Shell	^{54}Xe	^{36}Kr	^{18}Ar	^{10}Ne	^{2}He
$1s$	34 561	14 326	3203	867	25
$2s$	5 453	1 921	320	45	
$2p_{1/2}$	5 104	1 727	247		
$2p_{3/2}$	4 782	1 675	245	18	
$3s$	1 145	289	25		
$3p_{1/2}$	999	223			
$3p_{3/2}$	937	214	12		
$3d_{3/2}$	685				
		89			
$3d_{5/2}$	672				
$4s$	208	24			
$4p$	147	11			
$4d$	63				
$5s$	18				
$5p$	7				

structure constant. It follows from Eq. (2.6) that the electron energy in the ground state of the hydrogen atom is $\frac{1}{2}E_B$, which is used as a unit known as the rydberg (1 Ry $\approx$ 13.6 eV).

In atoms which have more than one electron the potential deviates from the Coulomb form and the degeneracy of levels in respect of the orbital quantum number is relieved. The spin–orbit interaction splits the levels with nonzero orbital momenta into doublets. This gives rise to a system of one-electron occupied levels typical for each atom, which is given in Table 2.1 for rare-gas atoms.

Each bound electron of the system is apparently at some energy level and, according to the Koopmans theorem, the ionization energy of a level is equal to the energy of a level taken with the opposite sign, i.e., it is equal to the absolute value frequently known as the binding energy of a level.

The nature of the wave function ψ depends on the field of forces which is experienced by an electron. In what is known as the valence approximation all the electrons of an atom can be divided in the ground state into two groups, those with the potential governed only by the nucleus and by the other electrons belonging to the same atom (core electrons), and those for which we have to allow also for the effects of the fields of the atoms in the environment (valence electrons).

The group of the core electrons may be regarded as including all the electrons of isolated atoms and also electrons of the inner shells of atoms in molecules and in the condensed state of matter. The group of the valence electrons includes electrons in the outer electron shells of nonisolated atoms, which give rise to chemical bonds and are largely transformed to the collective state in the field of all the nuclei in the system.

When we consider the wave functions of electrons in the final state ψ_f, it follows from the definition of the ionization process that we are dealing mainly with free or quasifree electrons, the wave functions of which are written in the form

$$\psi_f = C \exp(i\mathbf{qr}), \tag{2.10}$$

where $\mathbf{q}$ is the wave vector of an electron and C is a normalization factor which depends on the properties of space where the electron is located. For example, in the case of a homogeneous isotropic space we have $C = V^{-1/2}$, where V is the volume of the space, whereas in the case of a solid we have $C = V^{-1/2}u_q$, where u_q is a Bloch function.

In this (Born) approximation we ignore the interaction between an escaping electron and the object being ionized. This approximation does not give good results if the kinetic energy of the escaping electron is low, because then the wave function of this electron is distorted by its interaction with the ionization object.

We shall consider a photon as a plane monochromatic transverse electromagnetic wave in free space, characterized by the vector potential

$$\mathbf{A} = A_0\mathbf{n} \exp[i(\mathbf{kr} - \omega t)], \tag{2.11}$$

where A_0 is the wave amplitude; n is a unit polarization vector; $\mathbf{k}$ is the wave vector along the direction of propagation of a photon ($\mathbf{k} \perp n$, $k = 2\pi/\lambda$). The amplitude A_0 is related to the radiation intensity $I(\omega)$, i.e., to the density of the energy flux or to the absolute value of the Poynting vector averaged over an oscillation period:

$$I(\omega) = S\hbar\omega = \frac{cN\hbar\omega}{V} = \frac{\omega^2 A_0^2}{2\pi c}, \tag{2.12}$$

where S is the photon flux and N is the number of photons in a volume V.

The interaction of such a wave with an electron, which has a charge e and a mass m, is described by the following Hamiltonian provided the intensity is not too high:

$$H_{AB} = \frac{e}{mc}\mathbf{Ap}, \tag{2.13}$$

where $\mathbf{p}$ is the momentum of an electron in an electromagnetic field.

Using Eqs. (2.11)–(2.13) we find from Eq. (2.5) that the probability of absorption of a photon is

$$dW_{f0} = \frac{4\pi^2 e^2}{m^2\omega^2 c\hbar} I(\omega)|\langle f | \exp(i\mathbf{kr})\mathbf{np}|0\rangle|^2 d\rho(\hbar\omega), \tag{2.14}$$

where $d\rho(\hbar\omega)$ is an element of the density of states corresponding to the condition $E_f - E_0 = \hbar\omega$.

In specific calculations the momentum operator is usually represented in terms of the gradient or radius operator, using the identity

$$\langle f|\mathbf{p}|0\rangle \equiv -i\hbar\langle f|\mathbf{\nabla}|0\rangle \equiv i\omega m\langle f|\mathbf{r}|0\rangle.$$

It then follows from Eq. (2.14) that

$$dW_{f0} = \frac{4\pi^2\hbar e^2}{m^2\omega^2 c} I(\omega)|\langle f | \exp(i\mathbf{kr})\mathbf{n\nabla}|0\rangle|^2 d\rho(\hbar\omega) \tag{2.15}$$

or

$$dW_{f0} = \frac{4\pi^2 e^2}{\hbar c} I(\omega) |\langle f | \exp(i\mathbf{kr})\mathbf{nr} | 0 \rangle|^2 dp(\hbar\omega). \qquad (2.16)$$

The cross section of a transition deduced allowing for Eqs. (1.4), (2.15), and (2.16) is now

$$d\sigma_{f0} = \frac{dW_{f0}\hbar\omega}{I(\omega)} = \frac{4\pi^2\hbar^2 e^2}{m^2\omega c} |\langle f | \exp(i\mathbf{kr})\mathbf{n\nabla} | 0 \rangle|^2 dp(\hbar\omega) \qquad (2.17)$$

or

$$d\sigma_{f0} = \frac{4\pi^2 e^2\omega}{c} |\langle f | \exp(i\mathbf{kr})\mathbf{nr} | 0 \rangle|^2 dp(\hbar\omega). \qquad (2.18)$$

The form of the matrix element in Eqs. (2.15) and (2.17) is called the velocity form, whereas that given by Eqs. (2.16) and (2.18) is known as the radius form. If the wave functions are specified exactly, then both forms give the same result. However, in practice the wave functions used in this situation are more or less approximate. It is then necessary to bear in mind that the velocity form stresses the imprecision of the wave function in the region of an atomic nucleus, where it varies rapidly with the coordinates, and the radius form performs a similar role in the peripheral parts of an atom where the values of r are larger. In calculations it is more convenient to use the radius form because it does not contain derivatives. This form is also preferable in the frequently encountered case when the asymptotic behavior of the functions is well known.

In the case of ionization there is a free electron in the final state and ψ_f can be expressed in terms of Eq. (2.10), whereas for dp we find from Eq. (1.23) that (allowing for $\mathbf{p} = \hbar\mathbf{q}$)

$$dp = \frac{Vmq}{8\pi^3\hbar^3} d\Omega. \qquad (2.19)$$

It follows from Eq. (2.17) that

$$d\sigma_{f0} = \frac{e^2 q}{2\pi m\omega c} \left| \int \exp(i\mathbf{Kr})\mathbf{n\nabla}\psi_0 \, d\mathbf{r} \right|^2 d\Omega$$

$$= \frac{e^2 q(\mathbf{nq})^2}{2\pi mc\omega K} \left| \int \psi_0 \exp(-i\mathbf{Kr}) d\mathbf{r} \right|^2 d\Omega \qquad (2.20)$$

(after integration by parts allowing for the transverse nature of the field), where $\mathbf{K} = \mathbf{k} - \mathbf{q}$. It is clear from Eq. (2.20) that in the case considered the matrix element of the transition contains the Fourier transformation of the wave function of the ground state.

If before its interaction with a photon an atom is in the ground state, the function ψ_0 differs from zero only near the nucleus of this atom. Consequently, the matrix element of a transition also has nonzero values solely in the region comparable with a_B. Therefore, in the expressions given above it is meaningful to allow only for those values of r which satisfy $r \lesssim a_B$. At wavelengths of the incident radiation considerably greater than the dimensions of an atom, which corresponds to the energy of a photon less than a few kilo-electron-volts, we have $kr \approx ka_B \ll 1$ and we can use the expansion

$$\exp(i\mathbf{kr}) = 1 + i\mathbf{kr} + \tfrac{1}{2}(\mathbf{kr})^2 + \cdots \approx 1, \qquad (2.21)$$

which simplifies greatly Eqs. (2.15)–(2.18).

This is known as the dipole or long-wavelength approximation. One of its consequences is a selection rule for the orbital quantum number l: the matrix element of the transition differs from zero only when l changes by $\Delta l = \pm 1$ as a result of the transition. This rule permits the following transitions between the states: $s \leftrightarrow p$; $p \leftrightarrow s$; d; $d \leftrightarrow p$, f; and so on. Calculation shows that the probability of $l \rightarrow l + 1$ transitions exceeds the probability of the $l \rightarrow l - 1$ transitions by about an order of magnitude.

In the dipole approximation the photoabsorption cross section can be given a clear interpretation. Bearing in mind that $\mathbf{n \cdot r} = r \cos \theta$ (θ is the angle between $\mathbf{n}$ and $\mathbf{r}$) and averaging over the solid angle $(\overline{\cos^2 \theta} = \int_0^\pi \cos^2 \theta \sin \theta \, d\theta = \tfrac{1}{3})$, we find from Eq. (2.18) that

$$d\sigma_{f0} = \frac{4\pi^2 e^2 \omega}{3c} |\langle f | r | 0 \rangle|^2 d\rho(\hbar\omega) = \frac{4\pi^2}{3} \alpha r_{f0}^2 \hbar\omega \, d\rho(\hbar\omega). \qquad (2.22)$$

Here, $r_{f0} = \int \psi_f^* r \psi_0 \, dr$ is the matrix element of the transition radius. In the case when there is only one final state, we find that $\int d\rho(\hbar\omega) = (\hbar\omega)^{-1}$ and

$$\sigma_{f0} = \frac{4\pi^2}{3} \alpha r_{f0}^2 \approx 4\pi\alpha r_{f0}^2. \qquad (2.23)$$

It follows from Eq. (2.23) that the long-wavelength photon "sees" an atom as a circle of radius $2\alpha^{1/2} r_{f0}$. Such a "reduction" in the dimensions of an atom "seen" by a photon, compared with the geometric radius r_{f0}, is due to the weakness of the electromagnetic interaction characterized by the fine-structure constant $\alpha \approx \frac{1}{137}$.

If we introduce the concept of a dipole moment of a transition $d_{f0} = e r_{f0}$, Eq. (2.22) becomes

$$d\sigma_{f0} = \frac{4\pi^2 \omega}{3c} d_{f0}^2 \, d\rho(\hbar\omega). \qquad (2.24)$$

It is sometimes useful to describe a transition by a dimensionless quantity, known as the oscillator strength f_{f0} and defined by

$$f_{f0} = \frac{2m\omega}{3\hbar} r_{f0}^2 = \frac{2}{3} \frac{E_{f0}}{E_B} \frac{r_{f0}^2}{a_B^2}, \qquad (2.25)$$

where $E_{f0} = \hbar\omega$.

The sum of the oscillator strengths of transitions from a given state (or to a given state) is normalized:

$$\sum_j f_{ij} = 1 \text{ or } \int \frac{df_{ij}}{d\omega} d\omega = 1. \qquad (2.26)$$

Using Eq. (2.25), we find from Eq. (2.23) that

$$\sigma_{f0} = \frac{2\pi^2 e^2 \hbar}{mc} \frac{f_{f0}}{E_{f0}} \approx 1.1 \times 10^{-16} \, (\text{cm}^2 \, \text{eV}) \, \frac{f_{f0}}{E_{f0}}.$$

By way of example, we shall consider the photoionization of the hydrogen atom which is in the ground state. We then have

$$\psi_0 = (\pi a_B)^{-1/3} \exp\left(-\frac{r}{a_B}\right), \tag{2.27}$$

$$\int \psi_0 \exp(-i\mathbf{Kr}) d\mathbf{r} = (\pi a_B)^{-1/3} \int \exp\left(-\frac{r}{a_B} + iKr \cos\theta\right) 2\pi \sin\theta \, r^2 \, dr \, d\theta$$

$$= \frac{8\pi^{1/2} a_B^{3/2}}{(1 + a_B^2 K^2)^2}. \tag{2.28}$$

It follows from Eq. (2.20) that

$$d\sigma_{f0} = \frac{32 e^2 a_B^3 (\mathbf{nq})^2 q}{mc\omega(1 + a_B^2 K^2)^4} \, d\Omega. \tag{2.29}$$

If the energy of an absorbed photon is considerably higher than the ionization energy $(\hbar\omega \approx mv^2/2 = \hbar^2 q^2/2m \gg me^4/2\hbar^2$, where v is the velocity of a free electron), Eq. (2.29) is simplified. Then, on the one hand, we have $k/q = \omega/cq \approx mv^2/2\hbar cq = v/2c \ll 1$, i.e., $k \ll q$ and $K \approx q$, and on the other, we find that $q^2 \gg m^2 e^4/\hbar^4 = a_B^{-2}$, i.e., $q^2 a_B^2 \gg 1$. It follows from Eq. (2.29) that

$$\sigma \sim \frac{(\mathbf{nq}_1)^2}{a_B^5 (\hbar\omega)^{7/2}}, \tag{2.30}$$

where $\mathbf{q}_1$ is a unit vector in the direction of $\mathbf{q}$.

It is clear from Eq. (2.30) that the angular distribution of photoelectrons has a sharp maximum for $\mathbf{n} \| \mathbf{q}$, i.e., this maximum occurs along the direction of the electron vector of the incident photon and normally to the direction of incidence of the photon. A more detailed analysis of Eq. (2.29) shows that an increase in the photon energy $\hbar\omega$ shifts this maximum toward the direction of incidence of the photon which at relativistic electron energies $(v \to c)$ becomes aligned along this direction.

It also follows from Eq. (2.30) that the photoionization cross section decreases rapidly on increase in the photon energy [in proportion to $(\hbar\omega)^{-7/2}$]. Physically this is due to the fact that an increase in q increases the number of times that the wave function of a free electron changes its phase in the region of localization of an atomic electron and this reduces the matrix element of the transition (Fig. 2.1).

If we now consider the ionization of the K shell of atoms with $Z > 1$, we find that we need to allow for the fact that, in accordance with Eq. (2.8), the average radius of this shell for an atom with a nuclear charge Ze amounts to a_B/Z. Making the substitution $a_B \to a_B/Z$ in Eqs. (2.27)–(2.29), we obtain $\sigma_Z = \sigma_H Z^5$, where σ_H is given by Eq. (2.29). Therefore, an increase in Z causes the atomic photoionization cross section to rise steeply.

An increase in the ionization efficiency on increase in Z and on reduction in $\hbar\omega$ can be understood on the basis of the following qualitative discussion. It follows

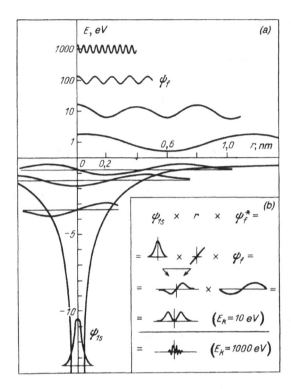

FIG. 2.1. Procedure used in calculation of a matrix element when dealing with the photoionization of the hydrogen atom: (a) wave functions of an electron in the initial and final states; (b) graphical representation of calculation of a matrix element for two values of the kinetic energy (10 and 100 eV) of a photoelectron.[118]

from the laws of conservation of energy and momentum that a free electron cannot absorb a photon. If an electron is in the field of a charge Ze, it becomes clear that an increase in Z causes the situation to deviate increasingly from the case of a free electron. Therefore, the ionization efficiency should increase. A similar effect is observed on reduction in $\hbar\omega$: in the case of a low-energy photon an electron can be regarded as more tightly bound to the nucleus.

It follows from the above discussion that the photoionization cross section of a many-electron atom can be described as follows. When $\hbar\omega$ becomes equal to the energy necessary to remove an electron bound least strongly to an atom, i.e., when it becomes equal to the first ionization potential of the atom, the photoionization cross section changes abruptly from zero to a value amounting to about $N\alpha a_B^2$ (N is the number of electrons in the outer electron shell of the atom). On further increase in $\hbar\omega$ the photoionization cross section decreases proportionally to $(\hbar\omega)^{-\beta}$, where β is a positive number of the order of several units (this dependence is a straight line when the coordinates $\log\sigma$ and $\log\hbar\omega$ are used). This occurs until $\hbar\omega$ becomes equal to the ionization energy of the next electron shell of the atom. When this happens, the cross section rises abruptly to its new value governed by the ionization probability of this shell. This rise is followed again by a fall of the cross section in accordance with a power law and two electron shells can now participate in the photon absorption. The pattern is repeated until the value of $\hbar\omega$ becomes much larger than the binding energy of electrons in the K shell of the investigated

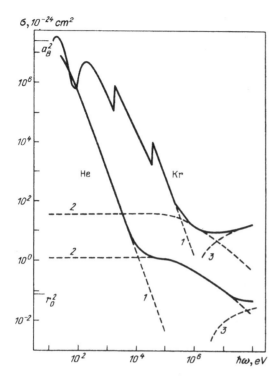

FIG. 2.2. Cross section for the inelastic interaction of He and Kr atoms with photons (continuous curves) and its components: (1) photoionization cross section;[107] (2) Compton scattering cross section calculated from Eq. (2.63); (3) cross section of the formation of electron–positron pairs in accordance with Eq. (2.39). Here, a_B is the Bohr radius and r_0 is the classical radius of an electron.

atom. On further increase in $\hbar\omega$, the ionization cross section vanishes. It follows from Fig. 2.2 that the experimental data for atoms with small and medium atomic numbers fit this picture well.

In the case of ionization of a subshell with a nonzero value of the orbital quantum number $l(l \geqslant 1)$ there may be deviations from the above simple relationships even in the one-electron approximation. One of the most important deviations is due to the fact that the effective potential U_{eff} of such electrons depends on l:

$$U_{\text{eff}} = U(r) + \frac{l(l+1)\hbar^2}{2mr^2},\tag{2.31}$$

where $U(r)$ is the Coulomb potential of the nucleus and of other electrons. This is an attractive potential and its value rises on increase in Z. The centrifugal potential is a classical analog of the second term in Eq. (2.31). For each value of $l \geqslant 1$ we have a range of Z where these two terms approximately balance each other out in a certain interval of the values of r. It is then found that $U_{\text{eff}}(r)$ has two valleys (one in the region of the nucleus and the other at the periphery of the atom) separated by a positive potential barrier. If the depth and width of the inner valley are insufficient for the formation of a bound state of an electron with a given l inside this valley, the electron in question is expelled to the periphery of the atom. Then, ψ_f has low values in the region of the nucleus where ψ_0 is localized and the probability of the $0 \to f$ transition at the ionization threshold is low. A considerable

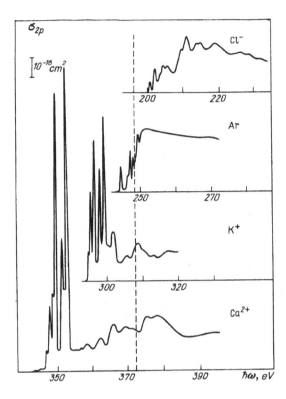

FIG. 2.3. Photoionization spectra of the 2p shell of an atom of Ar and of Cl^- (KCl), K^+ (KCl), and Ca^{2+} (CaF_2), ions made to fit the threshold ionization energy of this shell (dashed vertical line).[39]

overlap of the functions ψ_f and ψ_0 occurs only at the values of $\hbar\omega$ which exceed somewhat the threshold ionization energy. Consequently, the transition probability rises on increase in $\hbar\omega$ resulting in an apparent shift of the threshold toward higher energies. This is illustrated by the example of the 2p ionization spectra (representing $2p \rightarrow d$ transitions) of the Cl^- ion and the Ar atom shown in Fig. 2.3. For a given value of l an increase in Z deepens the internal potential valley, the barrier is lowered, and ψ_f goes over to the inner valley. The $l \rightarrow l + 1$ transition then has a high probability even below the ionization threshold. This effect is known as the electron collapse and it is clearly manifested in the case of the 2p ionization spectra of the K^+ and Ca^{2+} ions, which are also included in Fig. 2.3. It is thus found that the nature of the 2p absorption spectrum of isoelectronic argonlike ions changes rapidly on transition from Ar to K^+.

The one-electron approximation allows us to understand many characteristic features of the photoionization spectra of atoms. This applies particularly to light atoms and the inner electron shells of heavy atoms. However, in many cases the approximation is far too rough for a satisfactory description of the detailed structure of the spectra and the model has to allow also for the electron–electron interaction, i.e., for the many-electron effects.

The main many-electron effect in the ionization of the outer electron shells of the atoms is the dynamic polarization of the shell being ionized. As a result of such

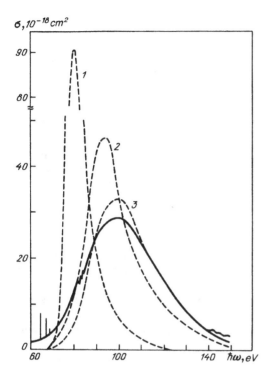

FIG. 2.4. Experimental (continuous curve)[89] and calculated[3,72] photoionization spectra of the 4*d* shell of an atom of Xe: (1) one-electron model; (2) Hartree–Fock model; (3) many-electron model.

polarization it is found that a photon interacts not with one electron, but with the whole shell to which this electron belongs. It seems that the shell resists the removal of an electron from itself. It is manifested by the fact that the ionization efficiency is reduced near the ionization threshold and the ionization spectrum shifts toward higher energies and expands.

In the case of ionization of deep electron shells of atoms the main many-electron effect is a modification of the outer electron shells during the time of escape of an electron from an atom. Such a modification results in an effective weakening of the field influencing a photoelectron (i.e., it reduces the effective value of *Z*) and, consequently, it suppresses the photoionization efficiency. An increase in the kinetic energy of a photoelectron makes this effect weaker and this shifts the maximum of the photoionization spectrum toward higher energies.

The influence of the many-electron effects on the photoionization spectrum is demonstrated by the example of the 4*d* absorption spectra of Xe shown in Fig. 2.4. A comparison of the experimental spectra with those calculated in the one-electron approximation without allowance for correlations (curve 1) shows that the process of ionization of the 4*d* shells is affected strongly by the many-electron effects. Allowance for the exchange interaction between electrons within the framework of the Hartree–Fock method improves somewhat the agreement between the experi-

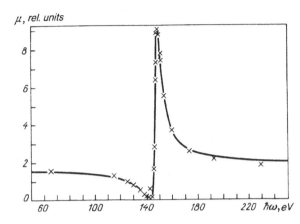

FIG. 2.5. Photoionization spectrum of the $4f$ shell of an atom of Gd (crosses) and its approximation on the basis of Eq. (2.32) using the parameters $\hbar\omega_m = 148.4$ eV; $\Gamma = 2.9$ eV; $q = 2.1$ (continuous curve).[86]

ment and theory (curve 2). A consistent allowance for the many-electron effects (curve 3) ensures a satisfactory agreement between the experimental and calculated spectra.

A frequently encountered case of the electron–electron interaction involves discrete excited states in one shell interacting with a continuous photoionization spectrum of another shell. This situation occurs every time that the energy of a discrete state exceeds the first ionization potential of an atom. This is true of all excitations of the inner electron shells because they are superimposed on the ionization continuum of an outer electron shell.

In such cases the wave function of the final state is a superposition of the wave functions of a discrete state (ψ_{fd}) and of a continuum (ψ_{fk}):

$$\psi_f = a\psi_{fd} + b\psi_{fk}.$$

As a result, a "discrete" electron acquires a "continuum" component, i.e., it acquires a kinetic energy and becomes delocalized. This process is called autoionization of the discrete state. The possibility of such an additional decay process shortens the lifetime of the discrete state and, consequently, broadens the corresponding spectral line. However, the interference between ψ_{fk} and ψ_{fd} may also distort strongly the profile of this line. In particular, in a certain part of the spectrum the total absorption may be below that in the surrounding continuous spectrum. In general, the total absorption in the region of a resonance (known as the Fano profile) can be described by the formula

$$\mu = \mu_A \frac{(q+\varepsilon)^2}{1+\varepsilon^2}, \tag{2.32}$$

where μ is the observed absorption coefficient; μ_A is the absorption coefficient which would have been obtained in the absence of discrete lines; ε is the energy separation from the maximum of the discrete line $\hbar\omega_m$ (in units of the half-width of this line Γ); q is a parameter which is a function of the strength of interaction between the discrete state and the continuum. A striking example of an experimental manifestation of the Fano profile is shown in Fig. 2.5.

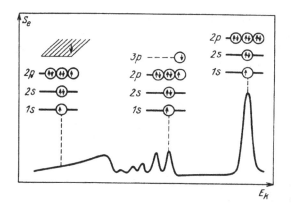

FIG. 2.6. Schematic representation of the electron spectrum in the course of the photoionization of the K shell of an atom of Ne showing the electron configurations of ions corresponding to characteristic details in the spectrum.[29]

One of the most interesting and important many-electron effects is the multiple ionization of atoms as a result of "one-electron" transitions. This process is possible because of the electron–electron interaction and its efficiency reflects the efficiency of this interaction. It is shown above that one of the most important conditions of the validity of the one-electron matrix element of the transition (2.5) is the assumption that Ψ_f and Ψ_0 refer to identical Hamiltonians, whereas, strictly speaking, Ψ_0 applies to the k-electron Hamiltonian and Ψ_f applies to the $(k-1)$-electron Hamiltonian (one electron is removed in the course of ionization). The probability that the atomic electrons represented in a nonionized atom by a set of quantum numbers n, l, s will retain the same quantum numbers after the ionization of an atom in accordance with Eq. (2.4) is

$$P_{n,\,l,s} = \left| \int \Psi_f^* \Psi_0 \, dr \right|^2 . \qquad (2.33)$$

If $P_{n,\,l,\,s} = 1$, then the approximation of Eq. (2.5) is rigorously valid. However, if $P_{n,\,l,\,s} \neq 1$, there is a probability, amounting to $1 - P_{n,\,l,\,s}$ that after removal of one electron from an atom there are also changes in the quantum numbers of the other electrons of the atom. Some of these electrons may then be transferred to the continuum, which corresponds to the multiple ionization of the atom. In other words, removal of an electron from an atom results in a sudden perturbation of the system of electrons of the atom and the system reacts to this perturbation by the creation of elementary excitations of the atom, i.e., by additional ionization. This is known as the shake-off of electrons. It is analogous to the shake-off of vibrations in the excitation of the electron subsystem in small molecules (see Sec. 1.2).

Figure 2.6 shows schematically the energy distribution of electrons emerging from the neon atom when its K shell is photoionized. The most energetic group of electrons has an energy $\hbar\omega - E_K$ (E_K is the ionization energy of the K shell, in this case of the Ne atom) and corresponds to the case when only one K electron is emitted from the atom and all other electrons conserve their quantum numbers. On the low-energy side this maximum is followed by one corresponding to the case when, in addition to the emission of one K electron, one further electron is shaken up from the $2p$ shell to the empty $3p$ shell. The subsequent maxima correspond to the shake-up of a $2p$ electron to higher bound states of the atom. Finally, the spectrum shows clearly the threshold corresponding to the emission not only of a K electron from the atom but also of a $2p$ electron with the minimum kinetic energy, i.e., corresponding to double ionization of the atom.

The shake-off of an electron from an atom may occur also when there is a sudden change in the state of its nucleus. This may be due to the radioactive decay of the nucleus, K capture, or even acquisition of a kinetic energy by the nucleus as a result of collision. If, for example, the nucleus acquires a velocity $\mathbf{v}'$ as a result of collision, then in terms of the coordinates in which the nucleus is at rest after the collision, the wave function of the ith electron of an atom acquires a translational motion component and is transformed in accordance with $\psi_0 \rightarrow \psi_0 \exp(i\mathbf{q}'\mathbf{r}_i)$, where $\mathbf{q}' = m\mathbf{v}'/\hbar$. The probability of the electron transition $0 \rightarrow f$ as a result of this event amounts, according to Eq. (2.33), to $P_{f0} = |\int \psi_f^* \exp(i\mathbf{q}'\mathbf{r}_i)\psi_0 \, d\mathbf{r}_i|^2$.

This mechanism may result in the ionization of light atoms by neutrons. For example, in the case of a head-on collision between a 1-MeV neutron and a ^{12}C nucleus the probability of ionization of the carbon atom is about 4% (Ref. 65).

We considered above the processes in which one photon participates and which are described by a first-order matrix element of Eq. (1.21). However, under certain conditions (high intensities or high photon energies) such processes are not very effective compared with those in which two photons participate: two-photon absorption (two photons before ionization, but none among the final products) or a photon is scattered (one photon before and after ionization). The probability of such processes also obeys the golden rule of Eq. (1.21), but instead of the first-order matrix element, it now contains a second-order matrix element

$$K_{f0} = \sum_k \frac{M_{fk} M_{k0}}{E_0 - E_k}, \tag{2.34}$$

where k represents an intermediate (virtual) state which is used to describe a two-photon process as a sequence of two one-photon processes represented by first-order matrix elements, and the index k is used to denumerate all the intermediate states. The principle of energy conservation then applies only to the overall two-photon process and not to its individual one-photon stages.

When matter is irradiated with strong laser radiation fluxes, an important role is played by two-photon or multiphoton absorption. This process is energetically possible if $N\hbar\omega \geqslant E_n$, where N is the number of participating photons and E_n is the binding energy of the nth level. When this condition is satisfied, the probability of N-photon ionization can be written as follows:

$$W_{f0}^{(N)} = \frac{mq}{4\pi^2\hbar} (2\pi\alpha S\omega)^N |K_{f0}^{(N)}|^2,$$

where $K^{(N)}$ is a matrix element of Nth order and

$$N\hbar\omega = E_n + \frac{\hbar^2 q^2}{2m}.$$

The selection rules for such a transition follow from the sum of the selection rules for all the participating transitions. Since a one-phonon transition is allowed in the case of different parities of the initial and final states, the process of absorp-

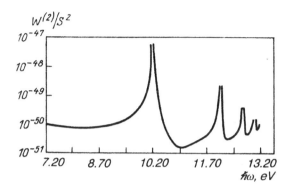

FIG. 2.7. Two-photon ionization spectrum of the hydrogen atom: the peaks represent terms of the Rydberg series in the one-photon absorption spectrum.[66]

tion of an even number of photons is allowed in the case when the parity of the participating states is the same.

We thus find the N-photon ionization threshold of an atom from the condition $N\hbar\omega = E_n$. We can expect the N-photon ionization cross section to be large if the energy $(N-1)\hbar\omega$ is equal to the energy of a transition to one of the discrete states of an atom. Figure 2.7 shows, by way of example, a part of the two-photon ionization spectrum of the hydrogen atom confirming this hypothesis.

The characteristics of the ionization and excitation of atoms by strong laser radiation beams have made it possible to develop highly sensitive selective methods for detecting a small number (in the limit just one) of atoms of a given type in a certain part of space.

The most reliable of these methods is known as resonant ionization spectroscopy. It can be described as follows[28,37]: The investigated region is irradiated with a laser tuned to the energy of a photon which, absorbed by a one- or a two-photon process, excites an atom of a given kind. The excited atom is ionized by the second (or third) photon from this or another laser (Fig. 2.8). A liberated electron is recorded with an electron counter. In view of a high (approaching unity) efficiency of detection of free electrons, it is possible to count practically every electron and, consequently, every ionized atom. Since the laser radiation is characterized by

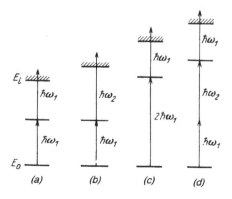

FIG. 2.8. Selective multiphoton ionization of atoms: (a) one photon excites an atom and another identical photon ionizes an excited state; (b) one photon excites an atom and another ionizes an excited state; (c) absorption of two identical photons excites an atom and a third identical photon ionizes it; (d) absorption of two different photons excites an atom and a third photon ionizes it.

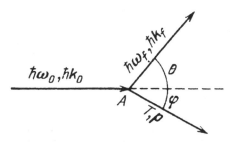

FIG. 2.9. Compton scattering of a photon by an atom of A.

narrow spectral lines and since the energy of the emitted photons can be controlled, the incident beam can be tuned in such a way that only the atoms of a given kind or of a specific isotope are ionized (and, therefore, detected) in the investigated region.

Electric fields can be used to drive ionized atoms out of the irradiated region. This makes it possible to use the method described above for isotope separation.

It should be pointed out that irradiation of atoms with high-intensity laser radiation can give rise to a further ionization process in the form of electron emission under the influence of the associated electric field. Atomic electrons subjected to high-intensity coherent photon fluxes experience strong electric fields which can result in the tunneling of these electrons to the states in the continuum. When electric fields are of the order of 10^8 V/cm, the binding energy of an electron is 15 eV, the thickness of the barrier is about 1.5 nm and its maximum height is about 9 eV, and the tunneling time is about 10^{-16} s.[45]

At high photon energies such that $\hbar\omega \gg E_K$, the photon-absorption cross section of an atom is small [see Eq. (2.30)] and the main inelastic photon–atom interaction process becomes the scattering of a photon by the atomic electrons, i.e., the Compton scattering. For these values of the photon energy $\hbar\omega$ we can ignore the energies of the bound electrons which behave as if they were free electrons at rest. The kinematics of such scattering follows from the laws of conservation of energy and momentum, which in general should be written down allowing for the relativistic energy of an electron after scattering (Fig. 2.9):

$$\hbar\omega_0 + mc^2 = \hbar\omega_f + (m^2c^4 + p^2c^2)^{1/2}, \tag{2.35a}$$

$$\hbar\mathbf{k}_0 = \hbar\mathbf{k}_f + \mathbf{p}, \tag{2.35b}$$

where p is the momentum of a recoil electron.

The system (2.35) can be used to calculate the energy lost by a photon and the equal kinetic energy T acquired by an electron:

$$\hbar(\omega_0 - \omega_f) = T = \frac{p^2}{2m} = \hbar\omega_0 \frac{\gamma(1 - \cos\theta)}{1 + \gamma(1 - \cos\theta)} = \hbar\omega_0 \frac{2\gamma}{1 + 2\gamma + (1 + \gamma)^2 \tan\varphi},$$

where $\gamma = \hbar\omega_0/mc^2$.

A calculation of the differential Compton scattering cross section, carried out using the second order of perturbation theory, gives the Klein–Nishina formula:

$$\frac{d\sigma}{d\Omega} = \frac{r_0^2}{2} \left(\frac{\hbar\omega_f}{\hbar\omega_0} \right) \left(\frac{\hbar\omega_0}{\hbar\omega_f} + \frac{\hbar\omega_f}{\hbar\omega_0} - \sin^2\theta \right), \qquad (2.36)$$

where $r_0 = e^2/mc^2 = 2.82 \times 10^{-13}$ cm is the classical radius of an electron. If $\gamma \gg 1$, it follows from Eq. (2.36) that

$$\sigma = \pi r_0^2 \frac{\ln 2\gamma}{\gamma}, \qquad (2.37)$$

whereas if $\gamma \ll 1$, we obtain

$$\sigma = \tfrac{8}{3} r_0^2 \approx 0.657 \times 10^{-24} \text{ cm}^2. \qquad (2.38)$$

The expression (2.38) represents the classical Thomson scattering cross section. We can see that in the course of the Compton scattering an electron acts as a solid sphere with a radius r_0. This is a consequence of the absence of the electron binding energy. The reason for the smallness of the maximum Compton scattering cross section ($\propto r_0^2$), compared with the maximum photoelectric absorption cross section ($\propto \alpha a_B^2 = r_0^2/\alpha^3$) also becomes clear: in the latter case a photon does not encounter a solid sphere but an electron cloud of size equal to an atomic orbit.

Since in the Compton scattering case each electron acts independently of other electrons, the cross section of this process in the case of an atom of atomic number Z is Z times greater than the cross section for the scattering by one electron. The cross section for the Compton scattering of photons by the atoms of He and Kr are plotted in Fig. 2.2.

There is currently much interest in the "inverse" Compton scattering of fast electrons by photons (which should not be confused with the backscattering of photons, which is also possible). This effect presumably accounts for the appearance of high-energy photons in cosmic rays. It is assumed that photons are generated by the scattering of electrons accelerated in cosmic magnetic fields by background radiation photons. However, this process can be used to generate high-energy photons also under terrestrial conditions. Scattering of accelerated electrons of energies 0.37–1.5 GeV by a laser beam of wavelength 514.5 nm has made it possible to generate quite strong (10^4–10^5 s^{-1}) beams of 5–78 MeV photons (Ref. 83).

If $\hbar\omega \geqslant 2mc^2$, one further interaction of photons with matter is possible: it is decay into an electron–positron pair in the field of nucleus. The presence of the latter ensures that the law of conservation of momentum is satisfied. The cross section of this process is

$$\sigma \approx \alpha Z^2 r_0^2 \gamma \ln 2\gamma; \qquad (2.39)$$

a comparison with Eq. (2.37) shows that the cross section differs by a factor of $Z\gamma^2\alpha$ from the Compton scattering cross section. The cross sections for the creation of pairs as a result of the interaction of photons of different energies with the atoms of He and Kr are also included in Fig. 2.2.

2.2. Photoionization of solids

The efficiency of the interaction of photons with large numbers of atoms in the condensed state and in particular with solids is best described using the concepts and quantities which follow from a phenomenological analysis of the interaction of electromagnetic radiation with continuous polarizable media. These quantities are the permittivity

$$\varepsilon = \varepsilon_1 + i\varepsilon_2 \tag{2.40}$$

and the complex refractive index

$$N = n + ik, \tag{2.41}$$

related by the Maxwell equations:

$$\varepsilon = N^2, \quad n^2 - k^2 = \varepsilon_1, \quad 2nk = \varepsilon_2. \tag{2.42}$$

Here, n is the refractive index and k is the extinction coefficient. These quantities represent the phase and attenuation of an electromagnetic wave in matter.

The relationship between the real and imaginary parts of the permittivity are given by the Kramers–Kronig dispersion relationship

$$\varepsilon_1(\omega_0) - 1 = \frac{2}{\pi} \int_0^\infty \frac{\omega \varepsilon_2(\omega)}{\omega^2 - \omega_0^2} \, d\omega. \tag{2.43}$$

These quantities can be determined experimentally from the measured reflection R and transmission T coefficients in the case of monochromatic radiation:

$$R = \frac{(n-1)^2 + k^2}{(n+1)^2 + k^2}, \quad T = (1-R)\exp\left(-\frac{4\pi kd}{\lambda}\right), \tag{2.44}$$

where d is the thickness of the object in the direction of propagation of such radiation.

The quantity

$$\mu = \frac{4\pi k}{\lambda} = \frac{2\omega k}{c} = \frac{\varepsilon_2 \omega}{cn} \tag{2.45}$$

is known as the absorption coefficient. It is related to the microscopic characteristics of matter by the expression

$$\mu(\hbar\omega) = N_0 \sigma(\hbar\omega) = \frac{N_0 W_{f0}(\hbar\omega)}{S}, \tag{2.46}$$

where N_0 is the concentration of atoms and c/n is the phase velocity of light in matter.

If $I(x)$ is the intensity of monoenergetic radiation at a depth x in matter, the linear losses of the intensity represent

$$-\frac{dI(x)}{dx} = \mu I(x), \tag{2.47}$$

which gives

$$I(x) = I_0 \exp(-\mu x), \qquad (2.48)$$

where I_0 is the intensity of the incident radiation on the surface of a substance.

Allowed one-electron, valence and free states in solids form continuous bands because of the interaction of atoms. The lowest-energy ionization process corresponds to a transition of an electron from the valence to the conduction band. The minimum photon energy necessary for this process is equal to the width of the upper band gap E_g. This gap amounts to several electron volts in the case of insulators and is less than 1 eV for the majority of semiconductors. Metals do not have such a band gap.

A theory of solids deals with one-electron states of band electrons which are regarded as plane waves with an amplitude $u(r)$ which is a function of the lattice period and is known as the Bloch function. The wave function of such an electron in the mth band can be written in the form

$$\psi_{mq}(\mathbf{r}) = V^{-1/2} u_{mq}(\mathbf{r}) \exp(i\mathbf{q}\mathbf{r}), \qquad (2.49)$$

where $\mathbf{q}$ is the wave vector of an electron ($\hbar\mathbf{q} = \mathbf{p}$ is its quasimomentum).

In contrast to free electrons the quasimomentum of an electron in a crystal cannot have any value, except those that allow for the possibility of propagation of an electron wave in the field of the periodic potential.

In the case of wave functions of the type described by Eq. (2.49) the probability of an electron transition from the band 0 to the band f can be deduced from Eq. (2.15):

$$dW_{f0} = \frac{4\pi^2 \hbar e^2}{m^2 \omega^2 c V} I(\omega) \left| \int u_{fq_f}^* \exp(-i\mathbf{q}_f \mathbf{r}) \exp(i\mathbf{k}\mathbf{r}) \mathbf{n} \nabla u_{0q_0} \exp(i\mathbf{q}_0 \mathbf{r}) d\mathbf{r} \right|^2 d\rho. \qquad (2.50)$$

The integral in Eq. (2.50) can be represented as follows:

$$\int \cdots = \int \exp[i(\mathbf{q}_0 + \mathbf{k} - \mathbf{q}_f)\mathbf{r}] u_{fq_f}^* \mathbf{n} \frac{\partial u_{0q_0}}{\partial \mathbf{r}} d\mathbf{r} + i\mathbf{q}_0 \mathbf{n} \int \exp[i(\mathbf{q}_0$$
$$+ \mathbf{k} - \mathbf{q}_f)\mathbf{r}] u_{fq_f}^* u_{0q_0} d\mathbf{r}. \qquad (2.51)$$

In view of the orthogonality of the Bloch functions belonging to different bands, the second term in Eq. (2.51) vanishes. The first term differs from zero only if

$$\mathbf{q}_0 + \mathbf{k} - \mathbf{q}_f = 0.$$

Since $k \approx \lambda^{-1}$ and $q_0 \approx q_f \approx a^{-1} \approx 10^7$ cm^{-1} (where a is the lattice constant), when the photon wavelength is $\lambda > 1$ nm, we can regard k as small compared with q, which leads to the following selection rule:

$$\mathbf{q}_0 = \mathbf{q}_f. \qquad (2.52)$$

This means that the electron quasimomentum is conserved in optical interband transitions in solids. The transitions satisfying this rule are known as direct. They are represented by vertical lines between the dispersion curves corresponding to different states $E(\mathbf{q})$ in Fig. 2.10.

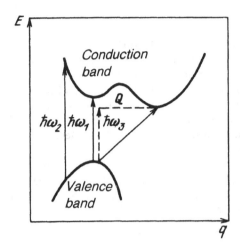

FIG. 2.10. Direct and indirect transitions between the valence and conduction bands. Here, photons $\hbar\omega_1$ and $\hbar\omega_2$ are absorbed as a result of direct transitions, whereas a photon $\hbar\omega_3$ is absorbed by an indirect transition accompanied by the absorption of a phonon with a quasimomentum $\hbar Q$.

It follows from Eqs. (2.50) and (2.51) that the probabilities of direct interband transitions are

$$dW_{f0} = \frac{4\pi^2 \hbar e^2}{m^2 \omega^2 cV} I(\omega) \left| \int u_{f q_f}^* \mathbf{n} \nabla u_{0 q_0} \, d\mathbf{r} \right|^2 dp_q(\hbar\omega), \qquad (2.53)$$

where $dp_q(\hbar\omega)$ is an element of the density of states such that transitions between them conserve $\mathbf{q}$ and have an energy $\hbar\omega$. This element depends on the structure of the energy bands of a specific solid. It should be pointed out that in the case of parabolic energy bands, i.e., when carriers are quasifree, the density of states is given by Eq. (1.23), which can be rewritten in the form

$$\rho(E) = \frac{V m^{*3/2}}{2^{1/2} \pi^2 \hbar^3} E^{1/2},$$

where m^* is the effective mass of quasiparticles and E is their energy in a band.

Figure 2.11 shows, by way of example, the energy band structure of a KBr crystal demonstrating the relationship between this structure and the interband absorption spectrum.

If the electron–phonon interaction is strong, an electron transition may be accompanied by the emission or absorption of one or more phonons. The selection rule of Eq. (2.52) then becomes

$$\mathbf{q}_0 = \mathbf{q}_f \pm \mathbf{Q}, \qquad (2.54)$$

where $\mathbf{Q}$ is the quasimomentum of a participating phonon. Such transitions are known as indirect (see Fig. 2.10). In view of the relative weakness of the electron–phonon interaction, the probability of such transitions is less than the probability of direct transitions. They are therefore manifested only in those parts of the absorption spectrum of a solid where there are no allowed direct transitions. They are in particular responsible for the initial part of the fundamental absorption spectrum when the maximum of the valence band and the minimum of the conduction band are located at different points in the Brillouin zone.

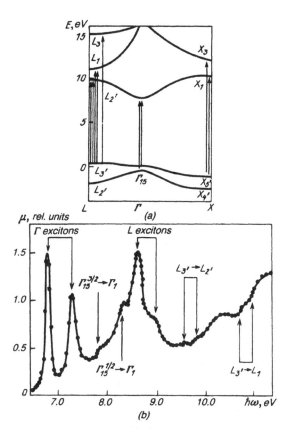

FIG. 2.11. Energy band structure of KBr (a) and the absorption spectrum[52] representing allowed direction transitions from the valence to the conduction band (b).

The selection rule of Eq. (2.52) typical of solids is in fact of limited validity. The most serious restriction follows from the circumstance that, because of the interaction with phonons and lattice imperfections (such as impurities or intrinsic defects), quasiparticles conserve a specific value of the wave vector only in a certain region in space and the size of this region can be represented by the mean free path l of these particles, which frequently amounts to 1–5 nm (see Secs. 2.4 and 2.5). The rule of Eq. (2.52) is meaningful only when the uncertainty of the wave vector Δq defined as $\Delta q \approx l^{-1}$ is less than the separation δq between different states in the E–q space, i.e., between different zones in the reduced zone scheme.

We shall estimate δq as a function of the electron energy E:

$$\delta q(E) = \frac{q_B}{s(E)\Delta E}, \qquad (2.55)$$

where q_B is the absolute value of the wave vector of an electron at the boundary of the Brillouin zone ($q_B = 2\pi/a$), ΔE is the average width of one energy band, and $s(E)$ is the number of bands per unit energy when the energy is E.

In the case of parabolic bands the density of states is proportional to $E^{1/2}$. Since the number of states in all the bands is the same, we can write down

$$s(E) = CE^{1/2},$$

where C is the normalization constant. Bearing in mind that $E_B = \hbar^2 q_B^2 / 2m^*$ is the energy width of the Brillouin zone, we find that

$$\int_0^{E_B} s(E)dE = 1 \quad \text{and} \quad s(E) = \frac{3}{2}\frac{E^{1/2}}{E_B^{3/2}}. \quad (2.56)$$

We then find that

$$\Delta E = \frac{\hbar^2}{2m^*}(|q_B + q|^2 - q^2) = \frac{\hbar^2}{2m^*}(q_B^2 + 2q_B q) \approx \frac{\hbar^2}{m^*}q_B q = \left(\frac{2\hbar^2}{m^*}\right)^{1/2}E^{1/2}q_B, \quad (2.57)$$

where it is assumed that $q \gg q_B$. Using Eqs. (2.56) and (2.57), we find from Eq. (2.55) that

$$\delta q(E) = \frac{2}{3}\left(\frac{m^*}{2\hbar^2}\right)^{1/2}\frac{E_B^{3/2}}{E} = \frac{\hbar^2}{6m^*}\frac{q_B^3}{E} = \frac{(2\pi)^3\hbar^2}{6m^* a^3 E}. \quad (2.58)$$

We can use Eq. (2.58) to determine the energy E_{cr} for which we have $\delta q = \Delta q$ and above which the law of conservation of the quasimomentum of an electron undergoing a transition loses its meaning. This gives the following estimate:

$$E_{cr} = \frac{8\pi^3\hbar^2 l}{6m^* a^3} \approx 10\text{--}100 \text{ eV}.$$

Photons with such energies belong to the vacuum ultraviolet part of the spectrum.

In the case of noncrystalline solids it is usual to assume that $l \approx a$ (in this case the symbol a denotes an interatomic distance), so that $\Delta q \approx q \approx a^{-1}$ and the selection rule of Eq. (2.52) is meaningless right from the beginning.

If the initial state of an electron participating in a transition does not belong to the valence band but to a deeper electron shell of some atom in a solid, the selection rule governing q again becomes meaningless: in this case ψ_0 cannot be regarded as a plane wave and the matrix element of the transition is not given by an expression of the type (2.51). The only practical valid rule is then the selection rule in respect of the orbital quantum number. Since in this case the function ψ_0 is localized in the region of atomic nuclei, the probability of a transition is described by the behavior of ψ_f also in the region of the nuclei, where this function retains largely the atomic nature, particularly when it represents a collapsed state (Sec. 2.1). Therefore, the general nature of the photoionization spectra of molecules in the x-ray region, where deep electron shells are ionized, is very similar to the corresponding spectra of atoms (Fig. 2.12).

In an analysis of the photoionization of solids we must also bear in mind that a photoelectron wave emerging from the region of localization of the atom being ionized is frequently reflected by the surrounding atoms and returns. Interference thus occurs between the emerging and reflected waves (Fig. 2.13) and the resultant amplitude and phase then depend on the photoelectron wave vector (i.e., on the incident photon energy) and the spectral dependence of the photoionization cross section becomes modified. We can allow for this effect by writing down the wave function of the final state in the form $\psi_f = \psi_f' + \psi_s$, where ψ_f' is the wave function of

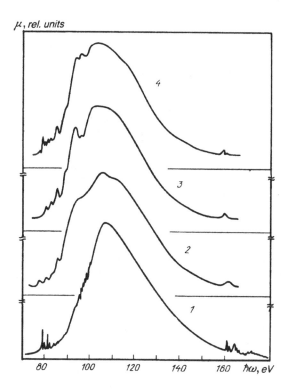

μ, rel. units

FIG. 2.12. Photoionization spectrum of the $4d$ shell of Cs atoms in various aggregate states: (1) free atom; (2) metal; (3) CsCl crystal; (4) CsCl molecule.[119]

the emerging electron without allowance for the environment and ψ_s is the change in this wave function as a result of the interference just described. The photoionization cross section can then be described in the form

$$\sigma = \sigma_0 + \sigma_s = \sigma_0[1 + \chi(\hbar\omega)],$$

where σ_0 is the atomic cross section and $\chi = \sigma_s/\sigma_0$ is the relative change in the cross section because of the interference.

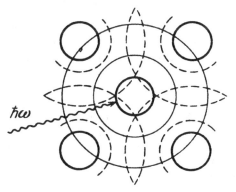

$\hbar\omega$

FIG. 2.13. Schematic representation of the structure of the standing waves formed by reflection, from neighboring atoms, of an electron wave emerging from the central atom ionized by a photon.

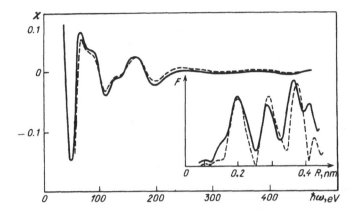

FIG. 2.14. Extended fine structure of the χ absorption spectrum of the 1s shell of Fe in deoxyhemoglobin and its Fourier transform $F(R)$ representing the distribution of the neighbors of an absorbing atom in respect of the distance R from this atom (the continuous curves are experimental and the dashed curves are theoretical[117]).

If the reflected wave amplitude is small, so that there is no need to allow for multiple reflection of electron waves, the quantity χ is described by

$$\chi(\hbar\omega) = \sum_j \frac{N_j |f_j(\hbar\omega)|}{qR_j^2} \exp(-2\mu|\mathbf{R}_j|) D_j(T) \sin[2\mathbf{q}\mathbf{R}_j + \alpha_j(\mathbf{q})].$$

(2.59)

Here, N_j is the number of neighbors of type j; $\mathbf{R}_j$ is the average distance from the absorbing atom to these neighbors; f_j and α_j are the amplitude and the phase shift due to the scattering of the electron wave; μ is a coefficient allowing for the attenuation of the electron wave in a crystal as a result of inelastic processes; $D_j(T)$ is a factor dependent on temperature T and allowing for the indeterminacy of $\mathbf{R}_j$ due to the lattice vibrations.

The resultant structure known as the extended x-ray absorption fine structure (EXAFS) is often observed clearly in the absorption spectra of solids at 50–1000 eV from the absorption edge. For fixed amplitudes and phase shifts of the scattering of an electron wave this structure depends only on the distances between the atoms. Therefore, the x-ray absorption spectra of solids determined over a wide part of the spectrum (which can be done using synchrotron radiation) provide an opportunity to determine with precision the interatomic distances in the investigated object since the Fourier transformation of the function $q\chi(\hbar\omega)$ to the real space is proportional to the distribution of N_j in terms of $\mathbf{R}_j$ (Fig. 2.14). This x-ray spectroscopic method for structural analysis is used widely at present.

In the case of condensed matter the one-electron approximation is fundamentally less accurate than in the case of atoms. In addition to the correlation of the motion of electrons in an atom being ionized, it is important to allow also for the correlation of the motion of electrons in the system of atoms. In some cases this can give rise to some important additional effects.

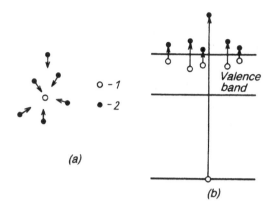

FIG. 2.15. Spatial (a) and energy (b) redistributions of the valence electrons in a metal as a result of the ionization of an inner electron shell of a metal atom: (1) holes; (2) electrons.

It is shown in Sec. 2.1 that if a one-electron transition alters considerably the potential of other electrons in an atom, an abrupt perturbation of this system of electrons may result in further ionization of the atom. In a solid such a perturbation can create excitations of the collective-state system of electrons, such as excitons, plasmons, and electron–hole pairs. For example, one-electron transitions in metals are accompanied by the appearance of low-energy electron–hole pairs near the Fermi surface (Fig. 2.15), which determines largely the profiles of the x-ray absorption edges.

An important factor which governs the photoionization cross section near the absorption edges is the interaction of a photoelectron with a resultant hole. This interaction may give rise to bound hole–electron states which correspond to discrete absorption bands near the edges (Fig. 2.12). Such states are known as excitons and in the case when a hole is located in an inner electron shell, they are called the core (x-ray) excitons.

The energy of an exciton state at the ith absorption edge is less than the energy of the edge itself and it is given by

$$E_{ie} = E_{ig} - \frac{E_B}{2\varepsilon^2} \frac{m_{ex}}{m},$$

where E_{ig} is the energy of the corresponding edge; $m_{ex}^{-1} = m_e^{-1} + m_h^{-1}$ is the reduced mass of an exciton; m_e and m_h are the effective masses of an electron and hole forming an exciton. The quantity $E_B m_{ex}/2m\varepsilon^2$ represents the binding energy of an exciton and it is sometimes called the exciton Rydberg [see Eq. (2.6)]. It is of the order of 1 eV for insulators and 10 meV for semiconductors.

2.3. Ionization by charged particles

The ionization of a physical object A by an electron can be regarded as a reaction

$$A + e \rightarrow A^{n+} + (n+1)e.$$

The simplest process of this type is single ionization

$$A + e \rightarrow A^+ + 2e, \tag{2.60}$$

sometimes denoted as the $(e, 2e)$ process. This notation reflects an important circumstance; after the ionization the electron causing it and that knocked out from the object being ionized are, in principle, indistinguishable.

A complete description of this process in the case of the simplest atom (that of hydrogen) requires solution of the secular Schrödinger equation for a system consisting of a proton and two electrons. However, even this simplest equation is far too complex to solve exactly (it represents a three-body problem!). Therefore, a whole series of approximations has been developed for solving problems of this kind. We shall use the simplest and most widely used of them, known as the Born approximation (see also Sec. 2.1). It should be pointed out that when a particle carrying a charge $Z_B e$ interacts with an atom with an atomic number Z_A, the condition of validity of this approximation is

$$\frac{Z_A Z_B e^2}{\hbar v_0} \ll 1, \qquad (2.61)$$

where v_0 is the velocity of the incident particle before the interaction.

The first simplification of the problem before applying the approximation involves allowance for the interaction of the incident electron with an atom only in a limited region of space. It is then assumed that outside this region the incident electron is a plane wave, whereas inside the region the atom is perturbed by the potential energy of the interaction of the incident electron with an atomic electron and the nucleus. The exact problem is thus replaced with one which can be tackled by perturbation theory. This approximation works well if after the collision of an electron with an atom one of the escaping electrons leaves the system at a high velocity. We can then say nominally that an incident electron crosses rapidly the region occupied by an atom and its momentum changes from $\hbar q_0$ to $\hbar q_f$, and the atom undergoes a transition from a state 0 to a state f under the influence of the perturbation.

In this formulation of the problem the probability of the interaction of an electron with an atom is given by

$$dW_{f0} = \frac{2\pi}{\hbar} \left| \int \Psi_f^* H_{AB} \Psi_0 \, dr \right|^2 d\rho(E), \qquad (2.62)$$

where the functions are described by

$$\Psi_0 = V^{-1/2} \exp(i q_0 r_1) \psi_0(r_2), \qquad (2.63a)$$

$$\Psi_f = V^{-1/2} \exp(i q_f r_1) \psi_f(r_2), \qquad (2.63b)$$

and the Hamiltonian of the interaction is given by

$$H_{AB} = \frac{e^2}{|r_1 - r_2|} - \frac{e^2}{|r_1|}, \qquad (2.64)$$

where r_1 and r_2 are the coordinates of the incident and atomic electrons, respectively.

Since in the case of a free electron we have $E = p^2/2m$, the density of states described by Eq. (1.23) is

$$dp(E) = Vm^2 v_f \, d\Omega (2\pi\hbar)^{-3}. \tag{2.65}$$

As a result of the interaction the wave vector of the incident electron changes by

$$\mathbf{K} = \mathbf{q}_0 - \mathbf{q}_f. \tag{2.66}$$

It follows from Eq. (2.66) that

$$K = (q_0^2 + q_f^2 - 2q_0 q_f \cos\theta)^{1/2}, \quad \frac{dK}{d\theta} = \frac{q_0 q_f \sin\theta}{K}. \tag{2.67}$$

Here, θ is the angle of deflection of the scattered electron which will be later taken to be the polar angle.

The matrix element in Eq. (2.62) is

$$M_{f0} = V^{-1} \int \exp(-i\mathbf{q}_f\mathbf{r}_1)\psi_f^*(\mathbf{r}_2) H_{AB} \exp(i\mathbf{q}_0\mathbf{r}_1)\psi_0(\mathbf{r}_2) d\mathbf{r}_1 \, d\mathbf{r}_2$$

$$= \frac{e^2}{V}\left[\int \frac{\exp(i\mathbf{K}\mathbf{r}_1)}{|\mathbf{r}_1 - \mathbf{r}_2|} \psi_0\psi_f^* \, d\mathbf{r}_1 \, d\mathbf{r}_2 - \int \frac{\exp(i\mathbf{K}\mathbf{r}_1)}{|\mathbf{r}_1|} \psi_0\psi_f^* \, d\mathbf{r}_1 \, d\mathbf{r}_2 \right]$$

$$= \frac{4\pi e^2}{VK^2} \left(\int \exp(i\mathbf{K}\mathbf{r}_2)\psi_0\psi_f^* \, d\mathbf{r}_2 - \int \psi_0\psi_f^* \, d\mathbf{r}_2 \right). \tag{2.68}$$

The last transformation is based on the Poisson relationship

$$\int \frac{\exp(i\mathbf{K}\mathbf{r}_1)}{|\mathbf{r}_1 - \mathbf{r}_2|} d\mathbf{r}_1 = \frac{4\pi}{K^2} \exp(iKr_2).$$

In the case of elastic scattering of an electron ($\psi_0 \equiv \psi_f$) we find from Eq. (2.68) that

$$M_{f0}^{\mathrm{el}} = \frac{4\pi e^2}{VK^2}\left(\int |\psi_0|^2 \exp(i\mathbf{K}\mathbf{r}_2) d\mathbf{r}_2 - \int |\psi_0|^2 \, d\mathbf{r}_2 \right) = \frac{4\pi e^2}{VK^2}(F-1), \tag{2.69}$$

where the quantity

$$F = \int |\psi_0|^2 \exp(i\mathbf{K}\mathbf{r}_2) d\mathbf{r}_2 \tag{2.70}$$

is known as the atomic form factor.

It follows from Eq. (2.69) that the cross section describing elastic scattering of electrons by an atom (or another object) carries information on the distribution of the charge in the ground state of the atom or object. The method of electron diffraction structure analysis is based on this circumstance.

We shall calculate the elastic scattering cross section using the expressions in Eq. (2.67) and the circumstance that in the case of incidence of one electron at a velocity v_0 the flux of such incident electrons is $v_0 V^{-1}$, where $v_0 = v_f$, $q_0 = q_f$, and $K = 2q_0 \sin(\theta/2)$, leading to

$$d\sigma^{el}_{f0} = \frac{dW_{f0}V}{v_0}$$

$$= \frac{4m^2 e^4}{\hbar^4 K^4} (F-1)^2 \, d\Omega$$

$$= \left(\frac{e^2}{2mv_0^2}\right)^2 \sin^{-4}\frac{\theta}{2} (F-1)^2 \, d\Omega. \qquad (2.71)$$

Using Eqs. (2.27) and (2.70), we find that in the case of the hydrogen atom

$$F = \left[1 + \left(\frac{Ka_B}{2}\right)^2\right]^{-2}.$$

When the transferred momentum is small ($Ka_B \ll 1$), the form factor is $F \approx 1 - K^2 a_B^2/2$, and the elastic scattering cross section is described by

$$\sigma^{el}_{f0} \approx 4\pi m^2 \left(\frac{ea_B}{\hbar}\right)^4 = 4\pi a_B^2, \qquad (2.72)$$

i.e., we find that this cross section is comparable with the geometric cross section of the atom (more exactly, it exceeds the latter by a factor of 4).

If the transferred momentum is large ($Ka_B \gg 1$), we find that $F \ll 1$ and

$$\frac{d\sigma^{el}_{f0}}{d\Omega} = \left(\frac{e^2}{2mv_0^2}\right)^2 \sin^{-4}\frac{\theta}{2}. \qquad (2.73)$$

The last expression is identical with the classical Rutherford formula (1.17). It can be rewritten in the form

$$\frac{d\sigma^{el}_{f0}}{d\Omega} = \frac{1}{4}\left(\frac{c}{v_0}\right)^4 r_0^2 \sin^{-4}\frac{\theta}{2}, \qquad (2.74)$$

which shows that the cross section for this interaction is comparable with the classical cross section of an electron.

We shall note here a very interesting circumstance: although the conditions of validity of the classical approximation (1.18) and of the Born approximation (2.61) are such as to exclude categorically the possibility that both sets of conditions are satisfied simultaneously, they nevertheless give the same result. Moreover, a rigorous quantum-mechanical calculation gives the same result in the range where $Z_A Z_B e^2/\hbar v_0 \approx 1$, when neither of these approximations is valid. This feature is typical only of the Coulomb potential and is the reason why the Born approximation gives good results even under the conditions when it should not work.

In the inelastic scattering case (when ψ_0 and ψ_f are orthogonal), it follows from Eq. (2.68) that

$$M_{f0} = \frac{4\pi e^2}{VK^2} \int \exp(i\mathbf{K}r_2)\psi_0\psi_f^* \, dr_2 = \frac{4\pi e^2}{VK^2} \langle f| \exp(i\mathbf{K}r_2)|0\rangle. \qquad (2.75)$$

We shall consider the case when the momentum transferred to an atom is small ($q_f \approx q_0$, $Ka_B \ll 1$), so that $\exp(iKr_2) \approx 1 + iKr_2$. In view of the orthogonality of ψ_0 and ψ_f, the first term of the expansion M_{f0} vanishes. The second term gives

$$M_{f0} = \frac{4\pi e^2 i}{VK^2} \langle f | \mathbf{K} r_2 | 0 \rangle = \frac{4\pi e^2 i}{VK} \langle f | r_2 | 0 \rangle, \tag{2.76}$$

where r_2 is the component of the vector r_2 along the direction $\mathbf{K}$. Using Eqs. (2.65) and (2.76), we find from Eq. (2.62) that

$$dW_{f0} = \frac{4m^2 v_f e^4}{V \hbar^4 K^2} |r_{2f0}|^2 \, d\Omega \tag{2.77}$$

and then, using Eq. (2.67) as well as the equality $v_f/v_0 = q_f/q_0$, we obtain

$$d\sigma_{f0} = \frac{dW_{f0} V}{v_0} = \frac{4m^2 e^4}{\hbar^4 K^2} \frac{v_f}{v_0} |r_{2f0}|^2 \, d\Omega = \frac{8\pi m^2 e^4}{\hbar^4 q_0^2} \frac{dK}{K} |r_{2f0}|^2,$$

which after integration yields

$$\sigma_{f0} = \frac{8\pi m^2 e^4}{\hbar^4 q_0^2} |r_{2f0}|^2 \ln \frac{K_{max}}{K_{min}}. \tag{2.78}$$

We shall now estimate the maximum and minimum values of the transferred momentum K_{max} and K_{min} in the $q_0 \approx q_f$ case:

$$K_{max} = q_0 + q_f \approx 2q_0. \tag{2.79}$$

Bearing in mind that

$$q_0^2 - q_f^2 = \frac{m^2(v_0^2 - v_f^2)}{\hbar^2} = \frac{2m\Delta E}{\hbar^2},$$

where ΔE is the change in the kinetic energy of the incident electron as a result of the scattering, we obtain

$$K_{min} = q_0 - q_f \approx \frac{m\Delta E}{q_0 \hbar^2}. \tag{2.80}$$

It follows from Eqs. (2.79) and (2.80) that

$$\frac{K_{max}}{K_{min}} = \frac{2q_0^2 \hbar^2}{m\Delta E} = \frac{2mv_0^2}{\Delta E} = \frac{4E_0}{\Delta E}.$$

The cross section of Eq. (2.78) is now

$$\sigma_{f0} = \frac{8\pi m^2 e^4}{\hbar^4 q_0^2} |r_{2f0}|^2 \ln \frac{4E_0}{\Delta E} = \frac{4\pi E_B}{E_0} |r_{2f0}|^2 \ln \frac{4E_0}{\Delta E} = 6\pi a_B^2 \frac{E_B^2}{E_0 \Delta E} f_{f0} \ln \frac{4E_0}{\Delta E}. \tag{2.81}$$

Here, E_0 is the kinetic energy of the incident electron and f_{f0} is the oscillator strength of the transition described by Eq. (2.22).

Equation (2.81) represents the cross section of a transition of an atom between two states, 0 and f, under the perturbing influence of the incident electron. In the ionization case that state f lies in the continuum and all the states with energies from 0 to E_f^{max} are accessible; here, E_f^{max} is determined by the requirement of energy conservation $E_f^{max} = E_0 - E_{nl}$, and E_{nl} is the binding energy of an electron with the quantum numbers n and l. In allowing for the participation in the process of ionization of atoms of all these states accessible from a fixed initial state nl, we must integrate the cross section σ_{f0} over all possible values of E_f and multiply the result by N_{nl}, the number of nl-electrons in an atom

$$\sigma_{nl} = N_{nl} \int_0^{E_f^{\max}} \frac{d\sigma_{f0}}{dE_f} \, dE_f. \tag{2.82}$$

Replacing ΔE with E_{nl} and using Eq. (2.26), we find from Eqs. (2.81) and (2.82) that

$$\sigma_{nl} = 6\pi a_B^2 \frac{E_B^2}{E_0 E_{nl}} N_{nl} \ln \frac{4E_0}{E_{nl}},$$

which is close to the frequently used Bethe formula:

$$\sigma_{nl} = \frac{4\pi e^4}{m v_0^2} \frac{N_{nl} b_{nl}}{E_{nl}} \ln \frac{4E_0}{E_{nl}}, \tag{2.83}$$

where

$$b_{nl} = E_{nl} \int_0^{E_f^{\max}} \frac{df_{f0}}{dE_f} \frac{dE_f}{E_f} = \frac{2}{3} a_B^{-2} \frac{E_{nl}}{E_B} \int_0^{E_f^{\max}} \frac{d|r_{2f0}|^2}{dE_f} \, dE_f$$

is a constant which, depending on Z, n, and l, has a value within the range 0.2–0.6; it is frequently assumed that $b_{nl} = 1$.

Comparing Eqs. (2.81) and (2.23), we can see, that in the Born approximation considered in the limit of small transferred momenta both in the case of a perturbation by the electric field of fast electrons and by an electromagnetic wave, the electron transitions in atoms are governed by the matrix element of the radius vector of the transition, i.e., for both types of radiation the same electron transitions occur in the atoms. The main difference between these two types of the ionization processes is as follows: (1) if the incident photon disappears as a result of this perturbation, then the incident electron is conserved, but it loses the kinetic energy equal to the energy of an electron transition in an atom; (2) in the case of electron irradiation the characteristic quantity is the geometric cross section of the atom being ionized ($|r_{2f0}| \approx a_B$), which is approximately $\alpha^{-1} \approx 137$ times larger than the cross section for the ionization of an atom by photons; (3) if the absorbed photon induces only transitions of energy $E_{f0} = \hbar\omega$, then an electron can induce all those transitions which are characterized by $E_{f0} \leqslant E_0 - E_{nl}$.

The Born approximation is invalid at low values of E_0. The energy of the incident electron is then only slightly higher than the ionization energy and after the scattering both electrons become slow, interference takes place between them, and during their motion within an atom the distribution of the other electrons in an atom changes. Relevant calculations show that near the ionization thresholds, where $E_0 - E_{nl} \ll E_{nl}$, we can use the relationship

$$\sigma_{nl} \sim (E_0 - E_{nl})^\gamma, \tag{2.84}$$

where the value of γ is close to unity ($\gamma \approx 1.127$). It follows from the experimental evidence that this relationship does indeed apply provided the separation from the

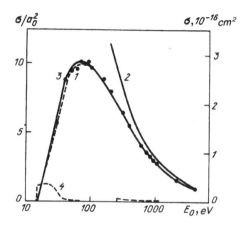

FIG. 2.16. Cross section for the ionization of Ar atoms by electrons: (1) experimental;[91] (2) calculated using the Born approximation; (3) calculated using Eq. (2.85), compared with the cross section of ionization by photons (4). Data taken from Ref. 107.

threshold does not exceed more than a few electron volts, but it may be strongly distorted by bound and autoionizing states.

It follows from the above discussion that the cross section σ_{nl} increases with the energy E_0 when this energy is low, but at high values of the energy it decreases. Consequently, the dependence $\sigma_{nl}(E_0)$ should have a maximum at moderate energies E_0. This maximum is usually located at energies exceeding severalfold E_{nl}. The cross section σ_{nl} for the ionization of outer electron shells of atoms has a maximum at energies 100–200 eV. The maximum value of σ_{nl} amounts to about πa_B^2 for the hydrogen atom and about $(0.2–0.3)\pi a_B^2$ per one electron in the case of heavier atoms. The Bethe formula of Eq. (2.83) describes well the experimental data in the range $E_0 > 200$ eV (Fig. 2.16).

In view of the difficulties encountered in a consistent derivation of an analytic expression for the cross section describing the ionization of atoms by electrons in a wide range of energies, it is quite usual to employ various semiempirical expressions. One of the most convenient and exact is the Lotz formula:[106]

$$\sigma_{nl} = a_{nl} N_{nl} \frac{\ln \dfrac{E_0}{E_{nl}}}{E_0 E_{nl}} \left\{ 1 - b_{nl} \exp\left[-c_{nl}\left(\frac{E_0}{E_{nl}} - 1\right) \right] \right\}, \qquad (2.85)$$

where a_{nl}, b_{nl}, and c_{nl} are the fitting constants. The good agreement between this formula and the experimental results is demonstrated in Fig. 2.16.

An important characteristic of the ionization process is the distribution of the kinetic energy of the electrons emerging from an atom. Clearly, if the Born approximation is valid, we can expect the appearance of two groups of electrons: fast "primary" and slow "atomic" electrons. Experiments show that in the majority of the events when the ionization is by fast electrons, such a distribution is indeed established. The maximum of the distribution of the slow electrons lies in the region of several electron volts and it shifts toward lower values on increase in E_0.

All this applies to nonrelativistic electrons ($E_0 \ll mc^2$). An allowance for the relativistic effects reduces formally to a slight modification of Eq. (2.81): all that is necessary is to make the replacement

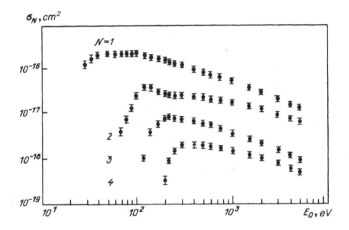

FIG. 2.17. Spectra of N-fold ionization of Cs^+ ions by electrons.[93]

$$\ln \frac{4E_0}{E_{nl}} \to \ln \left(\frac{4E_0}{E_{nl}(1-\beta^2)} \right) - \beta^2, \qquad (2.86)$$

where $\beta = v_0/c$.

In view of the long-range nature of the Coulomb perturbation, the one-electron description of the ionization by electrons represents the real situation much less satisfactorily than in the case of the ionization by photons. One of the many-electron effects is relatively effective multiple ionization of atoms irradiated with electrons (Fig. 2.17), in which case the cross section can be described by the empirical relationships

$$\sigma^{(N)} \approx 0.1\sigma^{(N-1)} \quad \text{when } E_0 \gg E^{(N)},$$

$$\sigma^{(N)} \approx c_N(E_0 - E^{(N)}), \quad c_N < c_{N-1} \quad \text{when } E_0 \gtrsim E^{(N)},$$

where $\sigma^{(N)}$ and $E^{(N)}$ are the cross section and the threshold energy for N-fold ionization.

In a quantitative analysis of one-electron transitions in solids induced by fast electrons it is usual, as in the case of atoms, to employ the Born approximation and the formulas that follow from it. When the inner electron shells are ionized, the results are naturally fully analogous to those obtained in the case of atoms. The main feature distinguishing solids from atoms is the presence of a large number of weakly bound valence electrons. An analysis of one-electron transitions of these electrons to the conduction band considered in the Born approximation yields Eq. (2.83) with the following modifications:

(1) N_{nl} is replaced with the number of the valence electrons per atom n_0;

(2) in the case of insulators and semiconductors the quantity E_{nl} is replaced by the band gap E_g, whereas in the case of metals it is replaced by the energy of the Fermi surface $E_F = \hbar^2 k_F^2/2m_e$ (k_F is the momentum of an electron on the Fermi surface);

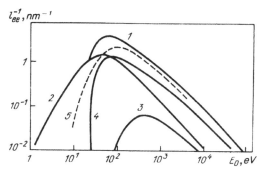

FIG. 2.18. Spectra representing the ionization of metallic aluminum by electrons (1), compared with the valence band of this metal (2), the 2p shell (3), the formation of plasmons in this shell (4), and the ionization spectrum of the valence band of Al_2O_3 (5). Data taken from Ref. 121.

(3) it is usual to omit b_{nl} (i.e., it is assumed that $b_{nl} = 1$). Then, the ionization cross section for the valence band is

$$\sigma_v = \frac{4\pi e^2 n_0}{m v^2 E_g} \ln \frac{2 m v^2}{E_g}. \tag{2.87}$$

An important result of the interaction of fast electrons with the valence electrons in solids is the excitation of collective excitations of the latter known as plasmons. Such oscillations appear because the additional field of the incident electron repels electrons in a solid and because of the inertia of their motion the initial displacement is greater than the equilibrium value. The system of electrons seems to attempt to liquidate the "extra" displacement and this gives rise to a transverse oscillation of frequency $\omega_{pl} = (4\pi n_0 e^2/m_e)^{1/2}$, which behaves as a quasiparticle of energy $E_{pl} = \hbar \omega_{pl}$. The cross section for the formation of a plasmon under the influence of a fast electron can be estimated from Eq. (2.87) by replacing E_g with E_{pl}.

Figure 2.18 shows, by way of example, the spectrum of the cross section for the ionization by electrons and its elementary components in the case of aluminum.

It follows from the above discussion that the energy losses due to the ionization by fast electrons interacting with matter are discrete and are due to the characteristic (of a given substance) electron transitions. This is manifested most strikingly in what are known as the spectra of the characteristic electron energy losses (CEEL) in solids. They represent the spectral dependences of the change in the energy of fast monoenergetic electrons crossing thin films of matter. They are very similar to the optical absorption spectra: both are governed by the same matrix elements. The main difference is that the bands due to the plasmon creation are clearly seen in the CEEL spectra and absent in the absorption spectra.

However, the CEEL spectra can provide additional information if we measure not only the energy loss, but also the deflection of an electron from its initial direction of motion, i.e., the transferred momentum [see Eq. (2.67)]. We can then plot the dependences of the energies of the created excitations on the momentum, i.e., we can plot the dispersion curves which are the most detailed characteristics of electron excitations in solids. This is not possible in the case of the absorption spectra of photons.

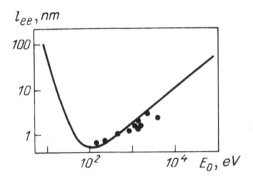

FIG. 2.19. Mean free path of electrons in the case of inelastic scattering by bound electrons in Al_2O_3 (the points are the experimental results).[121]

An important quantitative characteristic of the motion of fast electrons in matter is their mean free path l_{ee}, which is the average distance between two consecutive inelastic collisions with electrons in matter:

$$l_{ee}^{-1} = \sum_i \sigma_i N_i = \sum_i l_i^{-1}, \qquad (2.88)$$

where σ_i is the cross section of the ith elementary process; N_i is the density of the relevant electrons; l_i is the mean free path of the ith process. The summation in the above expression is carried out over all the elementary processes.

The dependence of l_{ee} on E_0 is described by a curve which is almost universal for all the solids and has a minimum in the region of $E_0 = 100$–300 eV where the value of l_{ee} is about 0.5–1 nm (Fig. 2.19). This minimum is due to the maxima of σ_v and σ_{pl} at the same values of E_0 (Fig. 2.18).

A rigorous quantitative analysis of the ionization of many-electron atoms by electrons and of the substances composed of such atoms is a difficult task. A semiempirical expression suitable for various utilitarian purposes can be obtained by introducing the concept of the average ionization potential of matter I_{av}. Its value is selected in such a way that if E_{nl} is replaced with I_{av} in Eq. (2.83), the resultant values of $\sigma \propto I_{av}^{-1} \ln(4E_0/I_{av})$ describe correctly the experimental data. It is found that the values of I_{av} obtained in this way for monatomic substances are approximately proportional to Z: $I_{av} \approx (10$–$13)Z$ electron volts. In the case of substances of complex composition the value of I_{av} is calculated from

$$\ln I_{av} = \sum_i f_i \ln I_{av,i}, \quad f_i = N_i Z_i \Big/ \sum_i N_i Z_i,$$

where N_i is the concentration of atoms characterized by $Z = Z_i$. For example, the average ionization potential of air is $I_{av} = 80.5$ eV.

In practical applications the ionization power of electrons in matter or, which is equivalent, the stopping power of a substance for electrons is usually described by two parameters: the linear energy losses and the range.

The linear energy losses are defined as $-(dE/dx)$, where E is the kinetic energy of electrons and dx is an element of their path. It follows from Eq. (2.83) that the linear losses due to nl electrons are

$$-\left(\frac{dE}{dx}\right)_{nl} \approx N_0 E_{nl}\sigma_{nl} = \frac{4\pi e^4}{mv^2} N_0 N_{nl} b_{nl} \ln \frac{2mv^2}{E_{nl}},$$

where N_0 is the concentration of those atoms which have the nl electrons. The linear losses are then

$$-\frac{dE}{dx} = \frac{4\pi e^4 N_0 Z}{mv^2} \ln \frac{2mv^2}{I_{av}} \approx \frac{4\pi e^4 N_A \rho Z}{mv^2 A} \ln \frac{2mv^2}{I_{av}} \approx K\frac{\rho}{E} \ln \frac{4E}{I_{av}}, \quad (2.89)$$

where $N_A = N_0 A/\rho$ is the Avogadro number; A and ρ are the values of the atomic weight (in atomic units) and the density of the substance; $K = 2\pi e^4 N_A \approx 2800$ keV^2 cm^2 g^{-1} is a constant which depends weakly on the nature of the substance.

The range $R(E_0)$ is defined by

$$R(E_0) = -\int_{E_0}^{0} \frac{dE}{dE/dx}. \quad (2.90)$$

Using Eq. (2.89), we obtain

$$R(E_0) = \frac{1}{\rho K} \int_0^{E_0} \frac{E\,dE}{\ln(4E/I_{av})} = \frac{I_{av}^2}{8\rho K} \int_0^{(2E_0/I_{av})^2} \frac{dy}{\ln y} = \frac{I_{av}^2}{8\rho K}\,\mathrm{li}\left(\frac{2E_0}{I_{av}}\right)^2$$

(li denotes the logarithmic integral). Since in the range $3 \leqslant y \leqslant 30$ we can assume that li $y^2 \approx y^{1.4}$, we estimate that

$$R(E_0) \approx \left(\frac{I_{av}}{2}\right)^{0.6} \frac{E_0^{1.4}}{2\rho K} = AE_0^{1.4}, \quad (2.91)$$

where A is a characteristic constant of a given substance.

The relationship (2.91) is satisfied well at low incident electron energies. For example, in the case of Si at $E_0 < 1$ keV, we have

$$R(\mathrm{nm}) \approx 32E_0^{1.4} \quad (\mathrm{keV}).$$

Other empirical expressions relating R and E_0 are also used frequently. At high values of E_0, we find that

$$R = AE_0 - B,$$

where A and B are constants which depend on E_0 and Z.

In the case of relativistic electrons ($E_0 \gtrsim mc^2 \approx 0.5$ MeV) it is necessary to allow also for other energy loss mechanisms, not directly related to the ionization. The main are the losses due to bremsstrahlung and Cherenkov radiation.

Inclusion of the bremsstrahlung contribution in the overall balance of the electron energy losses is important if $E_0 > 1600mc^2 Z^{-1}$, when estimates should be made using

$$\frac{(dE/dx)_{rad}}{(dE/dx)_{ion}} \approx \frac{E_0 Z}{1600mc^2}. \quad (2.92)$$

The energy losses due to the Cherenkov (also called Vavilov–Cherenkov) radiation represent a small fraction of all the losses experienced by relativistic electrons. A detailed allowance for such radiation losses is essential only when the radiation itself is of interest.

The ionization by heavy charged particles of mass $M \gg m$ (such as protons, α particles, multiply ionized atoms, etc.) can be described in the Born approximation by analogy with the ionization by electrons considered above. All that is necessary is to replace e^2 with $Z'e^2$, which corresponds to the interaction of a particle of charge $Z'e$ with an electron, and to replace the reduced mass of the system of two electrons $m/2$ with the reduced mass of a particle–electron system which is $mM/(m + M) \approx m$. Therefore, in the case of heavy charged particles Eq. (2.83) becomes

$$\sigma_{nl} = \frac{2\pi Z'^2 e^4 c_{nl} N_{nl}}{m v_0^2 E_{nl}} \ln \frac{2 m v_0^2}{E_{nl}}, \qquad (2.93)$$

where v_0 is the velocity of the incident particle and c_{nl} is a constant.

Other expressions given above are modified in a similar manner.

Thus, the ionization power of a heavy charged particle is Z'^2 times greater than the ionization power of an electron of the same velocity (but not with the same kinetic energy!).

When substances are ionized by protons, the cross section has a maximum at $E_0 \approx 100$ keV and its value is estimated to be 10^{-16}–10^{-15} cm^2. In solids with moderate values of the atomic number the energy losses experienced by such protons exceed 300 MeV/cm. If creation of one electron excitation in matter requires on the average 30 eV, then such ionization creates about 10^7 excitations per 1 cm. It means that every second or third atom along the path of a proton is ionized or excited. In the case of motion of an α particle with the same energy the linear energy losses in matter are approximately 10 times higher. Therefore, when solids are irradiated with heavy charged particles, a practically complete ionized and excited region of matter appears along their path and it is known as the track of a particle.

Since the laws of conservation of energy and momentum do not permit the transfer of large energies between colliding particles with very different masses [see Eq. (3.2)], the electrons knocked out from a substance as a result of the interaction with heavy charged particles are dominated by relatively slow electrons with a kinetic energy less than 100 eV (and with the maximum of the distribution at 10–30 eV).

An important feature of the motion of heavy charged particles in matter is the circumstance that, on the one hand, they may capture electrons from matter and, on the other, they can lose additional electrons. This effect is known as charge transfer and it is particularly important in the case of multiply charged ions. A quantum-mechanical analysis of charge transfer yields the following rule which can be deduced even intuitively: a fast ion captures electrons only to those orbits where the velocity of their motion v_n [Eq. (2.7)] is higher than the velocity of the ion itself and, conversely, in the interaction with matter it loses the electrons with velocities less than the velocity of the ion itself. As the velocity of an ion decreases in the course of its interaction with matter, there is also a reduction in its positive charge.

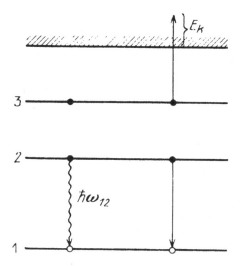

FIG. 2.20. Radiative and nonradiative transitions in an atom with an ionized inner electron shell.

This effect can be allowed for in Eq. (2.93) by regarding the parameter Z' as some effective value averaged over the whole path.

If a fast atom or an ion is incident on a thin film, it can lose a large number of its electrons in crossing the film and it can emerge from the film almost completely "bare." This effect is currently being used to generate highly charged ions for research purposes.

A set of colliding atoms exchanging electrons can be regarded as a short-lived molecule. When the relative velocity of atoms is high, such a molecule exists for a short time of the order of a/v (a is the diameter of the interaction region and v is the relative velocity). In spite of this, modern experimental techniques can be used to detect the appearance of such molecules using, for example, some characteristic features of the x-ray emission spectra.

If one of the colliding atoms is in an excited state, it can drop to the ground state by excitation or ionization of another atom. This process is known as the Penning ionization and it is frequently used in studies of the ionization of atoms and molecules at a given value of the transferred energy.

2.4. Electron–electron relaxation

After an ionization event an object is in a nonequilibrium state and begins to relax toward a new equilibrium. In general, such relaxation occurs by a multitude of elementary processes (Sec. 1.1). An atom with an ionized inner shell relaxes either emitting a photon or an electron (Auger process), as shown in Fig. 2.20. A free and/or quasifree electron can lose its energy by emitting photons or by scattering on electrons or phonons. The resultant secondary photons may be absorbed or scattered by analogy with primary photons.

Quantum mechanics regards the emission of a photon by an atom as a radiative transition from an excited to the ground state, i.e., as a process which is the reverse of the absorption of a photon. In quantitative estimates the probability of this process W'_{f0} can be obtained using the results of Sec. 2.1. Using the expression (2.16) in the dipole approximation (2.21), we find that

$$dW'_{f0} = \frac{4\pi^2 e^2}{\hbar c} I(\omega) |\langle f|\mathbf{n}\mathbf{r}|0\rangle|^2 dp(E_f). \tag{2.94}$$

Since in this process the final state involves creation of a free photon, the quantity $dp(E_f)$ represents the density of states of photons of energy $E_f = \hbar\omega$. Introducing $p = \hbar\omega/c$ and $dp/dE = c^{-1}$, we find from Eq. (1.23) that

$$dp(E_f) = \frac{V\omega^2}{(2\pi c)^3 \hbar} d\Omega. \tag{2.95}$$

Equations (2.94) and (2.95) yield

$$dW'_{f0} = \frac{\omega^3 e^2}{2\pi c^3 \hbar} |\langle f|\mathbf{n}\mathbf{r}|0\rangle|^2 d\Omega, \tag{2.96}$$

which allows for the fact that $I(\omega) = c\hbar\omega V^{-1}$.

In integration of this expression over angles we shall allow for the fact that $\mathbf{n}\cdot\mathbf{r} = \pi/2 - \mathbf{k}\cdot\mathbf{r}$ and $\cos(\mathbf{n}\cdot\mathbf{r}) = \sin(\mathbf{k}\cdot\mathbf{r}) = \sin\theta$. Since $\mathbf{n}\cdot\mathbf{r} = \mathbf{r}\cos(\mathbf{n}\cdot\mathbf{r})$, we obtain

$$|\langle f|\mathbf{n}\mathbf{r}|0\rangle| = \langle f|\mathbf{r}|0\rangle\sin\theta.$$

Then, integration of Eq. (2.96) gives

$$W'_{f0} = \frac{4\omega^3 e^2}{3\hbar c^3} |\langle f|\mathbf{r}|0\rangle|^2 = \frac{4\omega^3}{3\hbar c^3} d^2_{f0}. \tag{2.97}$$

We shall now estimate the result obtained. On the basis of Eq. (2.6) we can assume that $\hbar\omega \approx \frac{1}{2}Z^2 E_B$. In the case of dipole-allowed transitions, we find that $|\langle f|\mathbf{r}|0\rangle| = \mathbf{r}_{f0} \approx a_B Z^{-1}$ (see Sec. 2.1). Then, Eq. (2.97) yields

$$W'_{f0} = \frac{4\omega^3 \alpha r^2_{f0}}{3c^2} \approx \frac{Z^4 \alpha^3}{6\tau_B} \approx Z^4 \times 3 \times 10^9 \ (\text{s}^{-1}), \tag{2.98}$$

which gives the following estimate for the radiative lifetime of an ionized electron shell of an atom: $\tau_r = (W'_{f0})^{-1} \approx 3 \times 10^{-10} Z^{-4}$ (s). The value of τ_r decreases rapidly on increase in Z (Fig. 2.21).

The energies of the transitions between inner electron shells of atoms are found in the x-ray part of the spectrum. They are practically independent of the chemical state of an atom. Therefore, the characteristic x-ray emission spectra can be used to identify uniquely the emitting atoms. This is the basis of x-ray spectroscopic analysis.

On the other hand, the x-ray emission spectra representing the results of recombination of the valence electrons with inner holes created by ionization carry information on the energy and spatial distributions of the valence electrons and are used widely in studies of the structure of the valence bands of solids. The dipole selection rules make it possible to select inner levels with different values of the quantum number l and in this way obtain information also on the symmetry of the valence states.

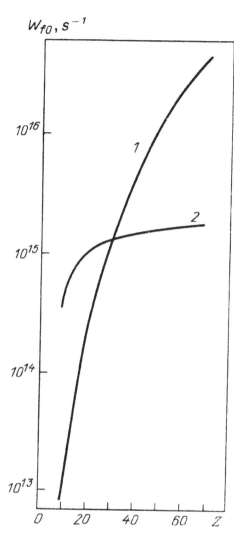

FIG. 2.21. Probability of occupancy of a hole in the 1s shell of an atom as a result of a radiative transition (1) and an Auger transition (2), plotted as a function of the atomic number (according to the data of Ref. 64).

Figure 2.22 provides a comparison, by way of example, of the emission spectra representing electron transitions from the valence band of a silicon crystal to the ionized 1s and 2p shells, on the one hand, and the function representing the density of the valence states found theoretically or deduced experimentally from x-ray electron spectra, on the other (Sec. 2.6). Clearly, the x-ray emission spectrum agrees well with the density of the valence states and it demonstrates that p-type states which have a high probability of being manifested in the 1s spectrum predominate near the top of the valence band (maxima C, D, E), whereas in the lower part of the band we are dealing with d-type and possibly s-type states which are manifested effectively in the 2p spectrum (maxima A and B).

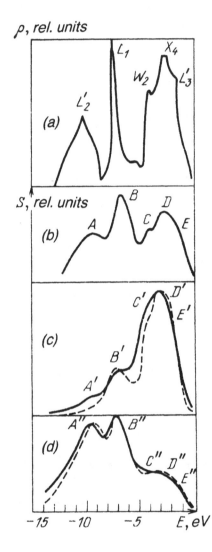

FIG. 2.22. Energy distribution of the density of states in the valence band of Si found theoretically (a) and from the x-ray electron spectra of Si (b). Zero energy corresponds to the top of the valence band. The bottom part of the figure shows the K (c) and $L_{2,3}$ (d) x-ray emission spectra.[41]

In interpretation of the real x-ray emission spectra we must bear in mind that the above pattern is only the first approximation to reality. We draw attention to two types of deviation.

Firstly, in addition to dipole radiation, we can expect also multipole radiation. For example, the probability of quadrupole transitions W_{f0}^{quad}, associated with the second term of the expansion (2.21), can be expressed in terms of the probability of dipole transitions W_{f0}^{dip} described by Eq. (2.94):

$$W_{f0}^{\text{quad}} \approx W_{f0}^{\text{dip}}(kr)^2 \approx W_{f0}^{\text{dip}}\frac{\omega^2 a_B^2}{c^2 Z^2} \approx \frac{1}{4} W_{f0}^{\text{dip}}\alpha^2 Z^2,$$

S, rel. units

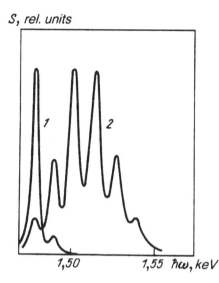

FIG. 2.23. X-ray K emission spectra of Al irradiated with protons (1) and with N^+ ions (2).[103]

where we are allowing for the fact that $k = \omega/c \approx Z^2 E_B/2\hbar c$. Therefore the relative probability of quadrupole radiation increases with Z as Z^2, and at high values of Z it can become quite significant.

Secondly, we must take into account the radiation from multiply ionized atoms, which appear in addition to singly ionized atoms (this is a particularly effective process when a substance is irradiated with charged particles). Roughly speaking the removal of each inner electron influences outer electrons in the same way as an increase in Z by unity. This corresponds to an increase in the energy of the transition and a shift of a given emission line toward higher energies. Such shifted x-ray lines are known as satellites. In the case of multiple ionization of an atom the spectrum has not only the main band, but also several satellite bands (Fig. 2.23).

Another effective process of decay of high-energy excitations is the Auger mechanism in which the transition energy is used to eject an additional electron from the system.

We shall consider a system of three discrete one-electron levels (with binding energies E_1, E_2, and E_3) and a continuum. We shall assume that the two higher levels are occupied and the lowest is vacant (see Fig. 2.20). The Auger emission from such a system is due to transitions $2 \rightarrow 1$ accompanied by the ejection of an electron 3 and $3 \rightarrow 1$ accompanied by the ejection of an electron 2 into the continuum. The kinetic energy of the ejected electron is in both cases the same: $E_k = E_1 - E_2 - E_3$. The process is possible if

$$E_k \geqslant 0. \tag{2.99}$$

The probability of such a transition can also be calculated using the golden rule of Eq. (1.22). In this case the perturbation energy is the energy of the electrostatic interaction between electrons undergoing transitions, i.e., $H_{AB} = e^2/|\mathbf{r}_{12}| \approx e^2/a_B = E_B$, where $\mathbf{r}_{12} = \mathbf{r}_1 - \mathbf{r}_2$. The wave functions considered in the one-electron approximation are

$$\psi_0(\mathbf{r}_1,\mathbf{r}_2)=\psi_2(\mathbf{r}_1)\psi_3(\mathbf{r}_2), \quad \psi_f(\mathbf{r}_1,\mathbf{r}_2)=\psi_1(\mathbf{r}_1)\psi_k(\mathbf{r}_2).$$

It then follows from Eq. (1.22) that

$$dW^a_{f0}=\frac{2\pi}{\hbar}\left|\int \psi_1^*(\mathbf{r}_1)\psi_k^*(\mathbf{r}_2)\frac{e^2}{|\mathbf{r}_{12}|}\psi_2(\mathbf{r}_1)\psi_3(\mathbf{r}_2)d\mathbf{r}_1\,d\mathbf{r}_2\right|^2 d\rho(E_f).$$
(2.100)

Hence, it is clear that from the microscopic theoretical point of view the Auger process is analogous to electron–electron scattering described above by Eqs. (2.62)–(2.64).

Calculation of integrals of the type represented by Eq. (2.100) in the hydrogenic approximation shows that the probability of an Auger transition in a given set of levels is independent of Z.

A more rigorous analysis of the Auger processes must allow also for the interaction of spins of electrons undergoing transitions with one another and with the orbital momenta. This lifts the degeneracy of the levels in respect of the corresponding quantum numbers and gives rise to several maxima in the spectrum of the kinetic energy of the Auger electrons. Moreover, experiments and relativistic calculations have shown that the probability of a given Auger transition rises slowly on increase in Z (see Fig. 2.21).

Allowing for the possibility of nonradiative transitions, we find that the relative probability of x-ray emission P_r can be described by

$$P_r=\frac{W^r_{f0}}{W^r_{f0}+W^a_{f0}}=\frac{1}{1+bZ^{-4}},$$
(2.101)

where b is a constant. This expression describes well the real situation at least in the case of radiative transitions involving the $1s$ shell. In this case the empirical value of b is 7.25×10^5 in the range $10\leqslant Z\leqslant17$ and 7.8×10^5 when $18\leqslant Z\leqslant54$.

At low transition energies the Auger process is always more likely than a radiative transition (see Fig. 2.21). This follows from a comparison of the relevant matrix elements in Eqs. (2.94) and (2.100). In the case of an Auger transition the perturbation energy is of the order of E_B, whereas in the case of a radiative transition the perturbation is $\alpha^{-1/2}$ times weaker. The situation is analogous to that considered earlier in comparing the effectiveness of the ionization of an atom by electrons and photons (Sec. 2.3). The situation changes at high transition energies. The difference between the values of the matrix elements of the transitions is then compensated or even outweighed by the difference between the densities of states: in the case of an Auger transition in the final state we have a free electron characterized by $\rho(E)\propto E^{1/2}$ [Eq. (2.65)], whereas in the case of a radiative transition we have a photon characterized by $\rho(E)\propto(\hbar\omega)^2$ [Eq. (2.95)].

The Auger electron spectrum, like the spectrum of the characteristic x-ray radiation, is largely specific to emitting atoms. This circumstance is used widely for the analysis of matter. Since in the condensed state the mean free path of an electron relative to inelastic interaction with other electrons in matter is only 1 nm (see Sec. 2.3), the unscattered electrons leaving a substance carry information mainly about the presence and state of atoms of a given kind on its surface. Auger spectroscopy is currently one of the most popular methods for the investigation of the surfaces of solids.

In view of the strong electrostatic interaction between electrons in one atom, many-electron effects play an important role in the Auger process. One of these effects is a multiple Auger transition: the energy liberated as a result of a nonradiative transition from an atom results in ejection of several electrons. For example, when a hole is filled in the 1s shell of Ne it is found that 8% of the total number of the Auger transitions involves the emission of two electrons.

Another manifestation of the many-electron effect is what is known as the postcollision interaction, which can be described as follows: if after the ionization of an atom, a photoelectron has a low kinetic energy, it moves away from an atom very slowly and may be captured by a bound state in the field of a new hole which appears as a result of the Auger decay of the primary hole. The energy which is then released is transferred to the Auger electron. If a photoelectron has a high kinetic energy this effect is no longer observed. Therefore, the Auger electron energy depends to some extent on the energy of an ionizing photon (or electron). Hence, it is clear that if the lifetime of a hole in any one of the inner electron shells of an atom is shorter than the time taken by a primary electron to move away from an atom, a discussion of the ionization process separately from the decay of the resultant excitation becomes physically meaningless. The ionization and relaxation of an atom should then be considered as a united process and this should be done on the basis of a many-electron treatment.

From the fundamental point of view it is of interest to consider two additional processes of filling of holes in the inner electron shells of atoms:

(1) a radiative Auger transition or a semi-Auger transition when one of the electrons in an outer shell fills radiatively an inner hole and at the same time a second electron changes its state in an atom or leaves it; in view of conservation of the total energy, the energy of an emitted photon is less than the difference between the energies of the initial and final levels;

(2) filling of two holes by two outer electrons accompanied by the emission of a photon whose energy is equal to the sum of the energies of the photons which would be emitted as a result of the corresponding one-electron transitions; for example, the emission spectrum of Ne includes a line corresponding to the $1s^0 2s^2 2p^m \rightarrow 1s^2 2s 2p^{m-1}$ transition.

In a quantitative sense these processes can be regarded as shake-off or shake-down of an additional electron as a result of the conventional radiative transition. The probabilities of these processes are very low and they are unimportant from the point of view of the balance of the intra-atomic processes. However, the occurrence of such processes in reality once again demonstrates limitations of the one-electron description of an atom.

One of the most urgent tasks in optics is currently the development of lasers emitting electromagnetic radiation with a wide spectral range, including far ultraviolet and x-ray ranges.[10] Such a laser can become feasible when a population inversion of x-ray levels can be established for a large set of atoms. The difficulty is the need for high pump powers because of the short lifetime of excited x-ray states and the high energy of x-ray photons. If the energy of the emitted photons is about

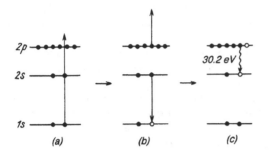

FIG. 2.24. Schematic diagram showing the proposed x-ray laser using the sodium ions:[76] the photoionization of the 1s shell of Na$^+$ (a) creates an Na^{2+} ion, which as a result of an Auger transition (b) goes over to the $1s^2 2s 2p^5$ configuration of the Na^{3+} ion, which then emits 30.2-eV photons as a result of the $2p \to 2s$ radiative transition.

10 keV, a typical lifetime of the states involved in the emission of these photons is 10^{-15} s and the required pump power is $\hbar\omega/\tau \approx 1$ W per atom. The only state of matter which can withstand such a high energy flux (and is consequently promising for lasing in the x-ray range) is a plasma.

The foremost promising source of such ultrahigh density of excitation of matter, which is needed in x-ray lasers, is a set of several high-power optical laser beams focused on one point. Although the necessary energy densities can be achieved in principle, there remains a fundamental difficulty: since the lifetime of excited atoms in the optical part of the spectrum is considerably longer than the lifetime of x-ray excitations, the latter begin to decay during the rise time of a pump pulse.

Attempts have frequently been made to increase the lifetime of x-ray excitations of atoms by employing atomic configurations in which the Auger processes are suppressed.[10] This can be achieved by removing the outer electrons from an atom. The desired results are most likely to be achieved as a result of $(n + 1)M \to nL$ transitions in ions of atoms with moderately large values of Z. It has been suggested that the following scheme may result in lasing of sodium ions: a flux of x-ray energy of about 1.1 keV ionizes the 1s shell of the Na$^+$ ions (which have the electronic configuration $1s^2 2s^2 2p^6$) converting them into Na^{2+} ions ($1s 2s^2 2p^6$); the Auger decay of the latter creates Na^{3+} ions ($1s^2 2s^2 2p^5$), which are characterized by a lifetime of 172 ps and undergo a radiative transition to the ($1s^2 2s^2 2p^4$) configuration emitting photons of 30.2 eV energy (Fig. 2.24). The necessary pump power is about 10^{12} W/cm^2.

A similar idea has already been put into practice in the case of selenium atoms: laser irradiation of selenium films created a population inversion between the $2p^5 3p$ and $2p^5 3s$ levels of the neonlike selenium, which resulted in considerable amplification of the emission lines at 20.63 and 20.96 nm (Ref. 109).

Consideration has been given also to the possibility of establishing a population inversion of nuclear levels and building of γ lasers or grasers.[19] The main advantage of γ lasers over those emitting in the x-ray range is that the lifetimes of excited nuclear states are frequently much longer than those of excited electronic states. However, the pump power is again extremely high. Synchrotron radiation is most likely to provide the necessary pump photon fluxes, whereas in the case of neutron pumping the necessary power might be achieved by a minor nuclear explosion.

It follows from this discussion that the relaxation of an atom with a hole in an inner electron shell occurs as a result of a cascade of radiative and Auger transitions

TABLE 2.2. Occupancy of holes in inner electron shells of the Ar atom.[70]

Initial hole	Final holes	Transition probability, 10^{13} s^{-1}	Relative transition probability, %
1s	2s2s	4.75	5.9
	2s2p	15.45	19.3
	2s3s	0.95	1.2
	2s3p	1.07	1.4
	2p2p	38.7	48.4
	2p3s	1.40	1.8
	2p3p	4.91	6.1
	3s3s	0.04	0.5
	3s3p	0.08	1.1
	3p3p	0.12	1.6
	2p	9.5	11.8
	3p	0.71	0.9
	Total	77.7	100
2s	2p3s	211	43.6
	2p3p	261	53.8
	3s3s	2.56	0.5
	3s3p	9.75	2.0
	3p3p	0.21	0.04
	3p	0.00	0.00
	Total	485	100
2p	3s3s	0.21	1.4
	3s3p	3.72	24.8
	3p3p	11.1	73.8
	3s	0.00	0.00
	Total	15	100

in a time 10^{-13}–10^{-15} s. Such transitions sometimes result in additional ionization of an atom as a result of shake-off of a number of electrons. The probabilities of specific transitions to the final charge state of an ionized atom depend on the actual shell which is ionized in the primary event. Tables 2.2 and 2.3 list the probabilities of various elementary transitions in an Ar atom and they also list the average charge state of a relaxed Ar atom after the ionization of different electron shells. It is worth noting the high rate of decay of a hole in the 2s subshell, which is due to a high probability of an Auger transition within one shell, known as the Koster–Kronig effect.

If a relaxing atom is surrounded by other atoms (as in molecules or condensed matter) the situation is somewhat different. In this case a relaxing atom may, in principle, capture electrons from neighbors. These are known as cross transitions. They appear sometimes in x-ray emission spectra and in Auger electron spectra. However, since the overlap of the wave functions of the inner electron shells of atoms is only slight, the probability of such processes is low. In many cases we can assume that an ionized atom first relaxes because of intra-atomic processes and then interacts with its neighbors. This is supported by experiments on molecules. For example, it has been shown that after the ionization of inner electron shells of iodine

TABLE 2.3. Distribution of argon ion charges (%) after relaxation of a hole in an inner electron shell.[69]

Ion charge	Shell from which electron is removed			
	K	L_1	$L_{2,3}$	M
1	0.7	0	0	85.5
2	10.5	2	74	13.5
3	7.8	72	24	1.0
4	42.7	24	2	
5	25.6	2		
6	10.3			
7	2.4			
8	0.6			

the CH_3I molecule decays in such a way as to suggest that the iodine atom undergoes first multiple ionization and then the molecule itself decays[71] (see also Sec. 3.3).

In the case of condensed media the secondary photons and electrons ejected from the primary ionization region interact with the medium surrounding this region and ionize it by the processes discussed in Secs. 2.1–2.3. The products of such secondary ionization events may result in tertiary ionization, and so on. Such a cascade ionization process continues until the resultant electrons and phonons are no longer able to cause further ionization. The final products of this process, which we shall call electron–electron relaxation, are low-energy electron excitations in a medium: conduction-band electrons, valence-band holes, excitons, and electrons and photons that have left a given object. The duration of such electron–electron relaxation is governed by the rate of the Auger and electron–electron scattering processes. If we assume that on the average one such relaxation event takes 10^{-15} s and that as a result of this event the energy of an electron excitation decreases by 10–100 eV, we find that the relaxation rate is 10^{16}–10^{17} eV/s, whereas the duration of electron–electron relaxation is estimated to be 10^{-15}–10^{-13} s, depending on the primary excitation energy.

During such relaxation there is a certain number of electrons and photons in a medium. Their distribution between the parameters of interest to us (such as the energy, spatial coordinates, or components of the momentum) is described by the kinetic equation in suitable coordinates and in an approximation acceptable for the problem in hand (see Sec. 1.3). We must bear in mind that one can frequently ignore radiative transitions and the propagation of secondary photons, because in the majority of cases the probability of decay of a hole in an inner electron shell of an atom by the Auger process is much higher than the probability of its radiative decay, and that fast electrons interact much more effectively with matter than photons of the same energy (see Sec. 2.3). Neglect of secondary photons is naturally unacceptable if a substance is irradiated with high-energy photons (γ rays) which interact with matter mainly via Compton scattering.

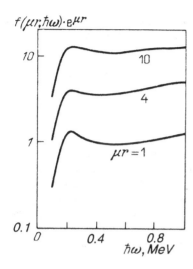

FIG. 2.25. Spectrum of radiation scattered from a unidirectional source of 1-MeV γ rays in iron (r is the distance from the front surface of a sample).[36]

Figures 2.25 and 2.26 give examples of the energy distribution functions of

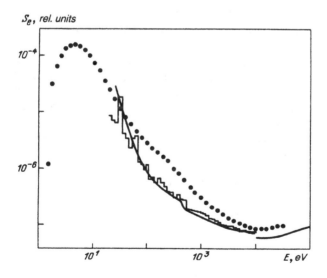

FIG. 2.26. Quasisteady spectrum of electrons in Al_2O_3 irradiated with β rays from ^{165}Dy with a maximum energy of 1.25 MeV (the points are the experimental results and the continuous curve together with the histogram represent theoretical calculations).[121]

secondary photons and electrons in an irradiated substance. The corresponding kinetic equations are a modification of Eq. (1.30) and have the form

$$\frac{df(E,t)}{dt} = G(E,t) + \int W(E+\Delta E, \Delta E)\, f(t, E+\Delta E)\, d\Delta E$$

$$- f(t,E) \int W(E, \Delta E)\, d\Delta E.$$

They can be solved by numerical methods, such as the Monte Carlo technique.

In many tasks in solid-state physics it is necessary to estimate the number of low-energy electron–hole pairs which appear in matter during the electron–electron relaxation stage. Since all the ionization processes finally create such pairs, we can assume that longer cascades of elementary relaxation processes result in averaging, so that the information on the nature of the primary excitation is lost and the resultant number of pairs N_{eh} is proportional to the energy ΔE_a transmitted to the investigated substance in the primary ionization events, i.e.,

$$N_{eh} = \Delta E_a / E_{eh}, \tag{2.102}$$

where E_{eh} is the average energy expended in creation in a given substance of one electron–hole pair incapable of further ionization. In the simplest variant we can assume that E_{eh} is the sum of the width of the band gap E_g, of the average kinetic energies of the final low-energy electron and hole E_{re} and E_{rh}, and of the energy E_v lost in collisions with phonons of those electrons and holes which are still capable of further ionization, i.e.,

$$E_{eh} = E_g + E_{re} + E_{rh} + E_v. \tag{2.103}$$

We shall estimate E_{eh} in the simplest case of the energy band structure with isotropic parabolic bands and with extrema at the point $q = 0$. We shall assume that the probability of filling, in the course of relaxation, of all the band states of electrons and holes is the same (random wave vector approximation). Then, the probability of creation of a carrier with an energy E is proportional to the density of states of this energy $\rho(E)$, so that we can calculate E_{re} as the average energy of electrons which at the moment of creation have an energy within the interval $0-E_t$, where E_t is the threshold energy for the ionization of a substance by an electron:

$$E_{re} = \int_0^{E_t} E\rho(E)\, dE \Big/ \int_0^{E_t} \rho(E)\, dE. \tag{2.104}$$

In the case of parabolic bands we have $\rho(E) \propto E^{1/2}$ (see Sec. 2.2). It then follows from Eq. (2.104) that

$$E_{re} = (\tfrac{3}{5}) E_t. \tag{2.105}$$

Similar assumptions give the same result also for E_{rh}.

In estimating E_t we begin by writing down the laws of conservation of energy and momentum in the case when the scattering of an incident electron of mass m_e and of velocity v_0 creates an electron of velocity v_e and a hole of velocity v_h and of mass m_h, and at the same time the velocity of the incident primary electron decreases to v_f. We then have

$$\tfrac{1}{2}m_e v_0^2 = E_g + \tfrac{1}{2}m_e v_f^2 + \tfrac{1}{2}m_e v_e^2 + \tfrac{1}{2}m_h v_h^2, \tag{2.106a}$$

$$m_e v_0 = m_e v_f + m_e v_e + m_h v_h. \tag{2.106b}$$

The system (2.106) does not have a clear general solution. Therefore, it is usually solved by invoking additional conditions. If we assume that E_t corresponds to the situation when the momenta of the three product particles are collinear and equal to one another, i.e., if we assume that $m_e v_f = m_e v_e = m_h v_h = \tfrac{1}{3}m_e v_0$, it follows from Eq. (2.106) that

$$E_t = (\tfrac{1}{2} m_e v_0^2)_{\min} = \frac{9E_g}{7 - m_e/m_h}, \tag{2.107}$$

which in the $m_e \approx m_h$ case gives

$$E_t \approx (\tfrac{3}{2})E_g, \tag{2.108}$$

whereas for $m_h \gg m_e$, we obtain

$$E_t \approx (\tfrac{9}{7})E_g. \tag{2.109}$$

In the case of a wide valence band if we assume that $m_h \approx m_e$, it follows from Eq. (2.103) subject to Eqs. (2.108) and (2.105) that

$$E_{eh} = E_g + 2 \times 0.9 E_g + E_v = 2.8 E_g + E_v. \tag{2.110}$$

For insulators the valence band is usually narrower than the band gap. In estimates we can assume that $E_g \gg E_{rh} \approx 0$ and $m_h \gg m_e$, so that Eqs. (2.103), (2.105), and (2.109) become

$$E_{eh} \approx E_g + 0.8 E_g + E_v = 1.8 E_g + E_v. \tag{2.111}$$

It is quite difficult to estimate E_v. A comparison of the measured values of E_{eh} with Eq. (2.110) shows that this energy is 0.5–1.0 eV. We can consequently assume that in the case of insulators we have $E_v \approx 0.2 E_g$ and $E_{eh} \approx 2 E_g$, whereas in the case of semiconductors we find that $E_v \approx E_g$ and $E_{eh} \approx (3-4) E_g$.

The available experimental values are in satisfactory agreement with these estimates (Fig. 2.27). Computer modeling of electron–electron relaxation in NaCl crystals irradiated with photons of energy $\hbar\omega = 20$–250 eV gives $0.54 E_g$ for E_{re} (on the assumption that $E_t = 1.5 E_g$), which is close to the estimate obtained from Eq. (2.105) (Ref. 100). The energy distribution of electrons in NaCl after such relaxation is plotted in Fig. 2.28 as a function of the incident photon energy. We can see that at high energies $\hbar\omega$ the distribution is readily averaged as a result of several elementary collision events and depends weakly on $\hbar\omega$. However, at low values of $\hbar\omega$ the distribution becomes a steep function of $\hbar\omega$, so that the procedure of averaging of Eq. (2.104) and of formulas such as Eqs. (2.109) and (2.110) becomes invalid. It should be pointed out that in the case of irradiation with photons of energy $\hbar\omega = E_g + E_t \approx (2-2.5) E_g$ the primary photoelectron becomes capable of creating a secondary electron–hole pair and the incident photon can produce two

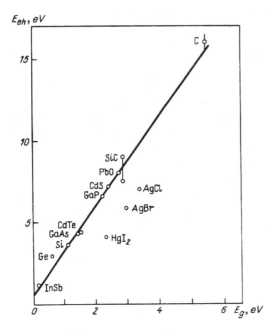

FIG. 2.27. Dependence of E_{eh} on E_g for semiconductors.[60]

electron–hole pairs. This phenomenon is observed in the excitation spectra of the luminescence emitted by ionic crystals and is known as photon multiplication.[25]

Interaction of positrons with matter represents a special case. The ionization interaction of positrons with matter is naturally quite analogous to the correspond-

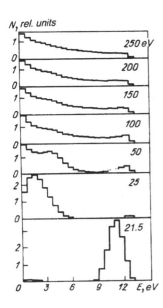

FIG. 2.28. Spectrum of the conduction electrons in an NaCl crystal after the electron–electron stage of relaxation at different incident photon energies.[100]

ing ionization interaction of electrons discussed above. The special features of the behavior of positrons follow from the possibility of their annihilation by the interaction with electrons, which results in photon emission.

It follows from the law of conservation of momentum that the presence of a fourth body (apart from an electron, a positron, and a photon) is essential in the case of such annihilation, because the fourth body accepts part of the momentum. This fourth body may be an atom to which an electron is bound (in this case we have a process which is the reverse of the creation of an electron–positron pair by a photon discussed in Sec. 2.1) or a second photon (in this case we have creation of two photons and the process is possible in the presence of a free electron). The second process is found to be much more effective, approximately by a factor of $(Z\alpha)^{-4}$, than the first. At nonrelativistic energies its cross section amounts to about πr_0^2 and the number of annihilation events per unit time, i.e., the rate of annihilation in matter characterized by an electron density ρ_e is

$$W_{ep} \approx \rho_e \pi r_0^2 c, \qquad (2.112)$$

which in the case of lead gives an estimate of $W_{ep} \approx 10^{10} \text{ s}^{-1}$.

Before annihilation a positron is very likely to become thermalized and form a bound system with some electron in matter, which is known as positronium. The latter is very similar to an atom of hydrogen except that the radius a_B is replaced with $2a_B$ (because instead of the reduced mass m we have $\frac{1}{2} m$).

The rate of spontaneous annihilation of positronium can be estimated using Eq. (2.112) where ρ_e is replaced with the density of the probability of finding an electron at the point where a positron is located, i.e.,

$$\rho_e \approx |\psi(0)|^2 = (8\pi a_B^3)^{-1} \approx 3 \times 10^{25} \text{ cm}^{-3},$$

which shows that in this case we also have $W_{ep} \approx 10^{10} \text{ s}^{-1}$.

The real positronium lifetime depends naturally on the total density of electrons in its vicinity. In a region in matter where the density of electrons is lower (for example, in the vicinity of positively charged defects or impurities in crystals—see Sec. 3.4) the positronium lifetime is longer. This makes it possible to investigate crystal defects experimentally by determination of the temporal and spatial distributions of the γ photons emitted from crystals irradiated with positrons. The experimentally determined positron lifetime in solids is 0.1–1.0 ps, which is in agreement with the above estimates.

Quasifree positronium migrates in a solid and the effective value of the diffusion coefficient is 0.1–0.2 cm^2/s, whereas the diffusion length is approximately 10^{-5} cm.

2.5. Relaxation of low-energy electron excitations in insulators and semiconductors

In addition to electron–electron relaxation of electron excitations in solids, these excitations interact also with the vibrations of the environment and the effect is known as the electron–phonon relaxation. In the case of excitations incapable of ionizing a substance, when the excitation energy is $E < E_t$ (we shall refer to these as low-energy excitations), the emission of phonons is the main energy relaxation mechanism in insulators and semiconductors.

In describing the effectiveness of the electron–phonon scattering by the golden rule of Eq. (1.21) the states 0 and f correspond to Bloch waves with different quasimomenta q_0 and q_f related by $q_0 - q_f = \pm Q$ in the case of first-order processes (Q is the phonon quasimomentum, the plus sign corresponds to phonon emission and the minus sign corresponds to phonon absorption). The Hamiltonian of such a perturbation $H_{ev} = D + e\varphi$ allows for the interaction of carriers with microscopic and macroscopic fields created by vibrations. The microscopic interaction is described by the deformation potential D, which represents a change in the energy of an extremum of a relevant energy band due to lattice deformation, whereas the macroscopic interaction can be represented by an effective electrostatic potential φ (which occurs in the macroscopic Maxwell equations) due to polarization vibrations. The first interaction predominates in the case of acoustic phonons, whereas the second in the case of optical phonons. The latter can be considered by analogy with an inelastic electron–electron scattering if we introduce the effective charge e^* and the polaron radius a_p:

$$e^* = \omega_{LO}\left(\frac{MV_0}{2\varepsilon^*}\right)^{1/2}$$

and

$$a_p = \left(\frac{\hbar}{2m^*\omega_{LO}}\right)^{1/2} \approx \left(\frac{a_B}{2\tau_B\omega_{LO}}\right)^{1/2},$$

so that $\varphi = e^*/a_p$ (here, ω_{LO} is the frequency of a longitudinal optical phonon; V_0 is the volume of a unit cell; $\varepsilon^{*-1} = \varepsilon_0^{-1} - \varepsilon_\infty^{-1}$; ε_0 and ε_∞ are the static and high-frequency permittivities, respectively). At a low carrier energy the polarization interaction is approximately $(M/m)^{1/2}$ times (where M is the mass of a lattice atom) stronger than the deformation interaction but the relative contribution of the latter increases on increase in the energy because of the short-range nature of the relevant potential. The polarization interaction predominates in ionic crystals if the carrier energy is less than 2–3 eV (Refs. 61 and 131). The average time τ_{ev} between two consecutive events of hot-electron scattering (relaxation time) is 10^{-14}—10^{-15} s (Fig. 2.29) which corresponds to the mean free path $l_{ev} = 1$–10 nm. In the case of substances with the predominant covalent or metallic type of binding the polarization interaction is weaker and the values of these parameters are somewhat larger.

The electron–phonon interaction results in thermalization of carriers after

$$n = \frac{(2N_v + 1)E_0}{\hbar\omega_v}$$

scattering events (for every $N_v + 1$ phonon emission events, there are N_v phonon absorption events); $N_v = [\exp(\hbar\omega_v/k_BT) - 1]^{-1}$ is the phonon occupation number; ω_v is the frequency of participating phonons.

The carrier thermalization time in ionic crystals irradiated with soft x rays is 10^{-11}–10^{-12} s. The process of thermalization can be described in terms of spatial coordinates quite satisfactorily as diffusion with a characteristic length $L = (D_e\tau_{ev})^{1/2} = (n/3)^{1/2}l_{ev}$, which in the case of ionic crystals amounts to $L \approx 10^2$–

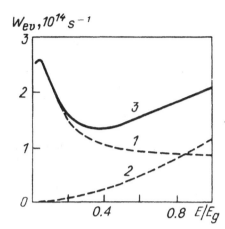

FIG. 2.29. Rate of energy relaxation of electrons in NaCl crystals as a result of the scattering by optical (1) and acoustic (2) phonons (curve 3 represents the combined relaxation rate) plotted as a function of the electron energy.[131]

10^3 nm, whereas for typical semiconductors it is $L > 10^3$ nm. The drift along the energy axis occurs at a rate of $\hbar\omega_v/\tau_{ev} \approx 10^{12}$–$10^{14}$ eV/s.

The scattering centers that limit migration of electron excitations are intrinsic or impurity defects present in crystals. The process of scattering of low-energy excitations by neutral point defects is in many respects analogous to the elastic scattering of electrons by hydrogen atoms (see Sec. 2.3). An approximate theory yields the following relationship for the scattering cross section in the case when $E < \frac{1}{4}E_1$ (E_1 is the ionization energy of defects):

$$\sigma_{es} \approx \frac{20a_1}{q}, \quad a_1 = \left(\frac{m}{m_e}\right)^2 \varepsilon_0 a_B,$$

which corresponds to a relaxation time τ_{es} defined by

$$\tau_{es}^{-1} = v\sigma_{es}N_s \approx 20 \left(\frac{m}{m_e}\right)^2 N_s\varepsilon_0 a_B\hbar m^{-1} \approx 10^{-7}N_s(\text{cm}^{-3})\varepsilon_0 \left(\frac{m}{m_e}\right)^2 (\text{s}^{-1}),$$

where N_s is the defect concentration. In the case of charged point defects it is reasonable to assume that the scattering is due to the Coulomb potential of the type (1.15) which leads to the Rutherford formula (1.17) for the scattering cross section.

The characteristic time for the scattering of electrons by dislocations τ_{ed} is usually estimated from

$$\tau_{ed}^{-1} = N_d v_\perp l_d,$$

where N_d is the dislocation density (cm^{-2}): $v_\perp$ is the component of the velocity of an electron excitation normally to the dislocation axis; l_d is a parameter with the dimensions of length. In the case of semiconductors, estimates give an empirical value $l_d \approx 10a_B \approx 0.5$ nm. At moderately high dislocation densities $N_d \lesssim 10^8$–10^9 cm^{-2} we have $\tau_{ed} \gg \tau_{ev}$ and the scattering by dislocations can be ignored.

The interaction of electron excitations with vibrations and lattice defects frequently results in localization of these excitations. A quantum-mechanical analysis of such processes is among the most difficult tasks in solid-state theory (particularly because of the participation of a large number of phonons and since the adiabatic approximation is frequently invalid).

The most universal of the carrier localization processes in insulators and semiconductors is the capture of carriers by lattice defects. The capture centers may be intrinsic or impurity defects. For example, vacancies in ionic crystals have an effective charge opposite to the charge of the ion they replace and, therefore, they act as effective carrier-capture centers (anion vacancies for electrons and cation vacancies for holes). Some impurity centers, for example, Tl^+ and Ag^+ ions, can capture both electrons and holes: $Tl^+ + e \rightarrow Tl^0$, $Tl^+ + h \rightarrow Tl^{2+}$. In the case of a typical semiconductor, such as germanium belonging to the group IV in the periodic system, group V impurities usually behave as electron donors, whereas group III impurities behave as acceptors.

The ability of a given center A present in a concentration N_A to capture carriers is described by the capture cross section σ_A and the lifetime of a carrier relative to such capture is

$$\tau_A^{-1} = v\sigma_A N_A, \tag{2.113}$$

where v is the velocity of the carriers being captured.

In estimating σ_A it is usual to consider just a phenomenological approach: an acceptable approximation for the kinetic equation is used to calculate the flux of carriers reaching a local center. If the capture center is regarded as an absolute black sphere of radius R_0 and all the carriers are postulated to have the average thermal velocity and the same mean free path l_{ev}, then the diffusion approximation applied to an electrically neutral center yields

$$\sigma_A = \frac{\pi R_0^2}{1 + \frac{3}{4} R_0 l_{ev}^{-1}} ; \tag{2.114}$$

R_0 is usually of the order of the geometric radius of a center, i.e., it is of the order of the lattice constant. In many cases we have $l_{ev} \gg R_0$ and σ_A is described by the gas-kinetic expression:

$$\sigma_A = \pi R_0^2. \tag{2.115}$$

In the opposite case when $l_{ev} \ll R_0$, we obtain the diffusion expression:

$$\sigma_A = \frac{4}{3}\pi R_0 l_{ev}, \quad \sigma_A v = 4\pi R_0 D_e, \tag{2.116}$$

where an allowance is made for the fact that $D_e = \frac{1}{3} v l_{ev}$.

We can estimate τ_A for a typical neutral center by assuming that $v \approx 10^7$ cm/s, $N_A \approx 10^{17}$ cm^{-3}, and $\sigma_A \approx 10^{-15}$ cm^2. We then obtain $\tau_A \approx 10^{-9}$ s.

In the case of charged centers, when the electrostatic interaction occurs between the centers and carriers, the radius R_0 in the above expressions must be replaced with the radius of this interaction R'. The latter can be estimated from the fact that the potential energy $U(R')$ of its interaction with a center is equal to the average thermal energy $\frac{3}{2} k_B T$ (where k_B is the Boltzmann constant and T is the absolute temperature):

$$U(R') = \frac{Z_A e^2}{\varepsilon_0 R'} = \frac{3}{2} k_B T,$$

where Z_A is the effective charge of the center. If $Z_A = 1$, then

$$R' \approx \frac{2}{3} \frac{e^2}{\varepsilon_0 k_B T}. \qquad (2.117)$$

It follows from Eq. (2.117) that at room temperature a typical value of R' amounts to about 10 nm. However, in most cases we have $R' > l_{ev}$ and σ_A can be estimated for charged centers using conveniently Eq. (2.116), where R_0 is replaced with R'. Clearly, typical values of σ_A for charged centers amount to 10^{-12}–10^{-13} cm^{-1}. The ratio of the cross sections for the capture of a carrier by charged and neutral centers should be of the order of $R' l_{ev}/R_0^2 \approx 10^2$–$10^3$, which is confirmed by the experimental results reported in Ref. 2.

In the case of narrow-gap semiconductors there are always many free carriers which screen partly the charged capture centers. Under these conditions the interaction of a center with a carrier can be described by the screened Coulomb potential:

$$U(r) = \frac{Z_A e^2}{\varepsilon_0 r} \exp\left(-\frac{r}{l_D}\right),$$

where $l_D = (\varepsilon_0 k_B T / 8\pi e^2 n_0)^{1/2}$ is the Debye screening length and n_0 is the free-carrier density. If $n_0 > 10^{17}$–10^{18} cm^{-3}, the value of l_D is less than R' found from Eq. (2.117) and the effective radius for the interaction of a charged center with carriers is in practice determined by l_D.

Many solids exhibit an interesting phenomenon of localization of electron excitations in a regular matrix, known as the self-localization. We can understand the causes of this phenomenon by adopting the Landau idea, where it demonstrated that a carrier in an ionic crystal creates around it a potential well because of the polarization of the environment, which increases considerably the effective mass of a carrier $(m^* \rightarrow M)$ and in the limit results in carrier localization. A detailed theoretical discussion of this range of topics has shown that the long-range polarization interaction of an electron with the crystal lattice creates what is known as a large-radius or simply large polaron with the effective mass m_p greater than the effective mass of the "bare" carrier m_e (or m_h):

$$m_p \approx m_e(1 + \tfrac{1}{6}\alpha_p),$$

where α_p is the electron–phonon interaction constant given by

$$\alpha_p = \left(\frac{m_e e^4}{2\varepsilon^{*2}\hbar^3\omega_{LO}}\right)^{1/2} = (2\varepsilon^{*2})^{-1/2}\left(\frac{m_e}{m}\right)^{1/2}\left(\frac{E_B}{\hbar\omega_{LO}}\right)^{1/2}.$$

In the case of the usual values of α_p (≈ 0.1–10) such a polaron continues its motion in a crystal (what is known as the polaron energy band, which is m_p/m_e times narrower than the band of a bare carrier) and does not become self-localized. Moreover, ultraheavy electrons with $m_p \approx (10^3$–$10^4)m$ encountered in some molecular crystals[127] are also mobile.

The self-localization of an excitation may result from its short-range deformation interaction with acoustic vibrations. If the lattice deformation by the excitation is sufficiently strong, there is an abrupt increase in its effective mass by several orders

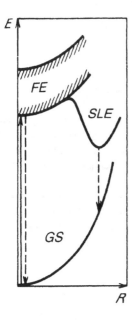

FIG. 2.30. Schematic representation of the exciton states: FE is the energy band of free excitons, SLE represents a self-localized exciton, and GS is the ground state of the investigated crystal. The continuous upward arrow represents the absorption of a photon and the downward dashed arrows represent the emission of a photon (E is the exciton energy and R is the configurational coordinate).

of magnitude and this results in self-localization of the excitation. However, this localization should be accompanied by an increase in the kinetic energy. In the most general form this follows from the uncertainty relationship $\Delta p \Delta x \approx \hbar$, which in this case means the following: if a band (Bloch) excitation is localized, the uncertainty of its coordinate decreases and this increases its momentum and kinetic energy. Such localization is possible if the interaction of a localizable excitation with the environment on reduction in its potential energy overcompensates the increase in the kinetic energy. Then, the self-localization process is related to overcoming of a potential barrier (Fig. 2.30). The additional interaction of a carrier with optical vibrations can then reduce the barrier height.

The self-localization of holes and excitons occurs in many ionic crystals.[38] A considerable lattice deformation then appears as a result of sharing of a hole or of the hole component of an exciton between two neighboring anions, which creates a two-center self-localized hole, representing an X_2^- molecule known as the V_k center, or an exciton $(X_2^{2-})^*$ (Fig. 2.31).

It has been established experimentally that free and self-localized states of an exciton are separated in alkali halide crystals by a potential barrier of height 10–20 meV, whereas in the case of holes such a barrier has not been found. The average time for the formation of self-localized holes in ionic crystals subjected to ionizing radiations is 10^{-11}–10^{-12} s.

The other type of self-localization of excitations is encountered in disordered (noncrystalline) solids, including glasses. Anderson showed that in any noncrystalline structure the lowest states in the conduction band (and the uppermost state in the valence band) are localized, because the distances between the equivalent lattice sites contributing to these states are large. On the energy scale there is a range occupied by such states (Fig. 2.32). The upper limit of this region determines

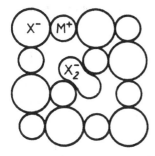

FIG. 2.31. Nuclear configuration of self-localized holes and excitons (X_2^-) in alkali halide crystals (X^- is an anion, M^+ is a cation).

the mobility edge. We can expect that at low temperatures all the thermalized carriers in such solids become localized. This process is known as the Anderson localization.

One further type of the localization of excitations, similar to the self-localization effect, is encountered in condensed media, mainly in liquids, consisting of polar molecules. This is solvation (the process in water is known as hydration) of an electron. In this case an electron is localized in the potential well formed as the result of polarization and/or reorientation of the surrounding molecules by its electric charge. In water such an electron (e_{eq}^-) represents a cloud of molecules with a radius of about 0.3 nm (Fig. 2.33).

The localization of an excitation lowers its energy by a certain amount E_l. In the case of the localization on defects the value of E_l depends on the distribution of the defect levels in the energy band scheme of a substance. In the self-localization process the energy E_l is governed by relaxation of the environment of the localizable excitation and is of the order of 0.01–0.1 eV for the Anderson localization in noncrystalline semiconductors and 0.1–1 eV in the case of the self-localization of

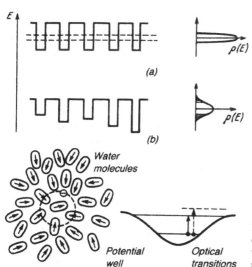

FIG. 2.32. Potential wells and density of carrier states in a crystal (a) and in a noncrystalline solid (b). In the latter case the edges of an energy band correspond to local states (shown shaded).

FIG. 2.33. Orientation of water molecules in the vicinity of a hydrated electron and optical transitions in the resultant potential well.[57]

holes in ionic crystals, but it rises to 1.8 eV for e_{eq}^-. The probability of liberation of localized carriers by thermal fluctuations is

$$P_l = P_{0l} \exp\left(-\frac{E_l}{k_B T}\right), \qquad (2.118)$$

where P_{0l} is a constant of the order of the vibration frequency of the environment ($\approx 10^{13}$ s^{-1}).

The final stage of the relaxation process of low-energy excitations is annihilation (excitons) and recombination (electrons and holes).

Free carriers recombine as a result of a transition of an electron from the conduction band to a vacant state in the valence band, i.e., a transition which is opposite to the photoionization of solids discussed in Sec. 2.2. The main selection rule for this process is the conservation of the quasimomentum of the electron undergoing the transition. This means that such recombination affects only those electrons and holes which have equal and opposite absolute quasimomenta. The high probability of the encounter of such carriers at the same points in space is likely only if the density of free carriers is high. In the case of a strong electron–phonon interaction one can expect also a phonon-assisted recombination process which is the reverse of indirect photoionization transitions.

Independently of the way and place of localization, a localized carrier is a recombination center for carriers of the opposite sign. The recombination cross section σ_r can be estimated from Eq. (2.114) or one of its modifications.

The recombination of low-energy electrons and holes created in one event occupies a special place among the recombination processes. It is known as the recombination of genetically related electrons and holes or twin recombination. This process is characterized by a high probability if $l_{ev} < R'$, when the probability of escape of a diffusing carrier outside the Coulomb attraction region of carriers of opposite sign is low. The efficiency of such recombination decreases on increase in temperature (because of a reduction in R') and on application of an electric field to an irradiated sample (this happens when oppositely charged carriers drift in opposite directions). The reduction in the effective density of carriers as a result of such recombination is frequently allowed for by introducing into an equation of the (1.36) type an effective carrier generation rate G_{eff}, which in the absence of an external electric field is given by

$$G_{eff} = G \exp(-R'/r_t),$$

where r_t is the distance in which a genetically related electron and hole become thermalized. It depends on l_{ev} and on the initial kinetic energy of carriers. The process of recombination of genetically related pairs of electrons and holes is important in ionic crystals if the initial kinetic energy of the relative motion of an electron and a hole does not exceed 1–2 eV (Refs. 2 and 5).

The recombination of carriers and the annihilation of excitons can be radiative or nonradiative. In the former case the liberated energy is emitted in the form of photons and is known as the recombination luminescence. If we apply the golden rule, we find that the problem is similar to radiative decay of x-ray excitations (Sec. 2.4). A considerable complication however arises from the circumstance that the states 0 and f are no longer described by the quasiatomic wave functions, but are formed under a strong influence of the crystal potential. Consequently, the value of W'_{f0} for processes of this type can vary within a wide range from 0 to 10^8 s^{-1}.

In the nonradiative recombination case the liberated energy can be released in the form of a packet of phonons or local vibrations, i.e., in the form of small displacements of a large number of atoms in the environment (i.e., heat) or large displacements of a small number of atoms (in the limit just one atom), producing structure defects (Sec. 3.3). When the spatial density of localized carriers is high, we can expect the Auger recombination process in which a carrier is ejected from a neighboring capture center as a result of the energy liberated at a given recombination center.

Closely spaced localized carriers with opposite charges can recombine by tunnel transitions. The probability of such transitions can also be estimated from Eq. (2.96). If the wave functions of the localized carriers are written in the usual exponential form of Eq. (2.27), the tunneling probability W_T can be found from Eq. (2.96):

$$W_T = W_{T0} \exp(-2R/a),$$

where R is the distance between the participating centers and a is the effective Bohr radius estimated to be $a \approx e^2/2\varepsilon_0 E_D \approx 0.1$–1 nm ($E_D$ is the donor ionization energy); here $W_{T0} \approx 100e^2\hbar\omega_T/m_e^2c^3a^2 \approx 10^7$–$10^8$ s^{-1} (Ref. 16).

The energy of a tunnel transition $\hbar\omega_T$ is governed by the gap between the energies of the levels participating before and after the transition. We must bear in mind that such a transition alters the charge state of the participating centers. Therefore, if the centers in question are initially in the neutral state, we find that

$$\hbar\omega_T = E_g - E_A - E_D + \frac{e^2}{\varepsilon_0 R}, \tag{2.119}$$

where E_A is the depth of the acceptor level in the band gap.

Localized (non-Bloch) excitations may migrate between close positions in a matrix. Such noncoherent migration is characterized by a jump probability P_t between configurations A and B, which in many cases can be estimated using an expression which is formally analogous to Eq. (2.118):

$$P_t = W_{AB} \exp\left(-\frac{\Delta_{AB}}{k_B T}\right). \tag{2.120}$$

The activation energy of a jump Δ_{AB} is governed by the difference between the excitation energies in the configurations A and B, by the barrier between these positions, and details of the electron–phonon interaction, whereas the preexponential factor W_{AB} is determined by the overlap of the excitation wave functions in the configurations A and B and by the mechanism of the interaction resulting in a jump (Coulomb or exchange).

We shall first consider the Coulomb interaction. In the case of allowed electron transitions the main role is played by the dipole interaction. If the energy and the wave functions of the states are labeled E_A and Ψ_A for the initial configuration and and E_B and Ψ_B for the final configuration, the probability of a transfer event deduced from the golden rule is

$$W_{AB} = \frac{2\pi}{\hbar} |\langle \Psi_A^* \Psi_B'^* | H_{AB} | \Psi_A' \Psi_B \rangle|^2 \delta(\Delta E), \qquad (2.121)$$

where $\Delta E = E_A' - E_A - (E_B' - E_B)$ and primes identify the excited states. The energy of the dipole–dipole interaction is

$$H_{AB} = \frac{\vec{d}_A \vec{d}_B}{R^3} - 3\frac{(\vec{d}_A \vec{R})(\vec{d}_B \vec{R})}{R^5} = \frac{\chi}{R^3} |\vec{d}_A| |\vec{d}_B|, \qquad (2.122)$$

where $\vec{d}_A = e\vec{r}_A$ and $\vec{d}_B = e\vec{r}_B$ are the dipole moments of the interacting systems; $\vec{r}_A$ and $\vec{r}_B$ are the radius vectors of the optical transitions; R is the distance between them; $\chi \approx \frac{2}{3}$ is a factor which depends on the angles governing the relative orientation of the configurations. The last equality in Eq. (2.122) is obtained by averaging over the angles on the assumption that all the mutual orientations of the dipoles are equiprobable and the polar axis is directed along R.

If the distance R is large, so that the dipole moments $\vec{d}_A$ and $\vec{d}_B$ as well as the states A and B, can be regarded as independent of one another (i.e., if there is no dynamic coupling between the initial and final configurations), we find from Eqs. (2.122) and (2.121) that

$$W_{AB} = \frac{2\pi\chi^2}{\hbar R^6} |\langle \Psi_A^* \Psi_B'^* | \vec{d}_A \vec{d}_B | \Psi_A^* \Psi_B \rangle|^2 \delta(\Delta E)$$

$$= \frac{2\pi\chi^2}{\hbar R^6} |\langle \Psi_A^* | \vec{d}_A | \Psi_A' \rangle|^2 |\langle \Psi_B'' | \vec{d}_B | \Psi_B \rangle|^2 \delta(\Delta E)$$

$$\sim \frac{1}{R^6} I_A(\omega) \mu_B(\omega).$$

The above expression follows from Eqs. (2.46), (2.24), and (2.97). Here, $I_A(\omega)$ is the emission spectrum of the configuration A; $\mu_B(\omega)$ is the absorption spectrum of the configuration B. An exact analysis gives what is known as the Förster formula:

$$W_{AB} = \frac{9\chi^2 c^4}{8\pi N_B \tau_A R^6} \int \frac{I_A(\omega)\mu_B(\omega)}{\omega^4} d\omega, \qquad (2.123)$$

where N_B is the concentration of the configurations B; τ_A is the lifetime of the configurations A relative to the spontaneous emission event.

The energy transfer process thus resembles a sequence of events consisting of the emission of a photon by an excited configuration A and the absorption of this photon by a configuration B which is in the ground state. In the case of hot transfer the function $I_A(\omega)$ represents the secondary radiation (including hot luminescence and scattering) spectrum, whereas τ_A is the lifetime of the relevant hot state.[50]

This applies to allowed dipole transitions between singlet states, i.e., to what is known as the singlet–singlet transfer. It follows from Eq. (2.123) that the probability of such a transition decreases proportionally to R^6 on increase in the distance. If the transition is forbidden, it may still be induced by the interaction of multipoles of higher order. For example, the dipole–quadrupole interaction results in transfer which is $(a/R)^2$ times less effective than the dipole–dipole interaction (where a is the radius of a localized excitation). The effectiveness of the energy transfer process decreases with distance proportionally to R^8.

If an excitation is in a triplet state, then the main energy transfer mechanism is the triplet–triplet process due to the exchange interaction between the configurations. The transfer probability is then governed by the overlap not of electric fields, but of the wave functions. It is described by the expression

$$W_{exch} = \frac{2\pi}{\hbar} P \int I_A(\omega) I_B(\omega) d\omega, \qquad (2.124)$$

where P depends on the overlap of the wave functions and in the asymptotic limit falls exponentially on increase in R. Such transfer may play an important role in molecular crystals where the distance between neighboring molecules is small. In view of the long lifetime of triplet states, we need to allow for the diffusion of an excited molecule when considering such an energy transfer mechanism. For example, if the lifetime of a triplet state is 10^{-3}–10^{-4} s, then even if the diffusion coefficient is relatively small (10^{-3}–10^{-4} cm^2/s), the length of a diffusion displacement $(D\tau)^{1/2}$ is considerable (1–10 μm) and it in fact determines the energy transfer distance.

It follows from the above discussion that the nature of the time dependences of the energy and spatial coordinates allows us to divide low-energy electron excitations into three classes: hot, thermalized, and localized.

The behavior of the whole system of excitations can in principle be described by writing down and solving the Fokker–Planck equation (1.34) or the Boltzmann equation (1.37) for each class of excitations. Since excitations are being continuously redistributed between these classes, the description of the system of excitations as a whole requires solution of a complex nonlinear system of equations which is possible only in the few simplest cases.

There are at present no sufficiently detailed experimental data on the behavior of the various classes of electrons, holes, and excitons, so that this approach is not very fruitful. It is therefore usual to employ what is known as the one-velocity approach in which it is assumed that all the excitations have a constant effective energy (or velocity). If the spatial distribution function of the excitations $f_e(\mathbf{r}, t)$ $= \int f_e(\mathbf{r}, E, t) dE$ changes little in the distance equal to the effective range of excitations l_{eff}, so that

$$\left| \frac{df_e(r,t)}{dr} \right| \ll \frac{f_e(r,t)}{l_{eff}},$$

then in the absence of external electric or magnetic fields the behavior of a system of excitations in space and time can be described by a diffusion equation of the (1.36) type:

$$\frac{\partial f_e}{\partial t} = D_e \frac{\partial^2 f_e}{\partial r^2} - \frac{f_e}{\tau_{eff}} + G_e, \qquad (2.125)$$

where τ_{eff} is the effective excitation lifetime and G_e is governed by the irradiation rate and the characteristics of the electron–electron relaxation stage.

If we ignore special cases of irradiation of solids with photons of certain resonance energies, then the main excitations that actually determine electronic properties of the irradiated substances are the conduction electrons and the valence holes. Since these quasiparticles exist simultaneously, but have different kinetic properties, the effective parameters in Eq. (2.125) are given by the expressions

$$D_{\text{eff}} = \frac{f_e + f_h}{f_e/D_h + f_h/D_e}, \quad \tau_{\text{eff}} = \tau_e \frac{\mu_e + \mu_h f_e/f_h}{\mu_h + \mu_e \tau_e/\tau_h}, \tag{2.126}$$

where μ_e and μ_h are the electron and hole mobilities. Equation (2.125) with the coefficients (2.126) is known as the ambipolar diffusion equation.

If the system of excitations can be regarded as spatially homogeneous, so that

$$\left| \frac{df_e(r,t)}{dr} \right| \ll \frac{f_e(r,t)}{L_{\text{eff}}}$$

[where $L_{\text{eff}} = (D_{\text{eff}}\tau_{\text{eff}})^{1/2}$ is the effective diffusion length of the active excitations], the diffusion term in Eq. (2.125) can be omitted and the spatial distribution function $f_e(r, t)$ of the excitations should be replaced with the concentration $N(t) = \int f_e(\mathbf{r}, t)dr$. Then, instead of Eq. (2.125) we obtain

$$\frac{dN}{dt} = G_e - \frac{N}{\tau_{\text{eff}}}, \tag{2.127}$$

and hence the steady-state concentration N_∞ of excitations is given by

$$N_\infty = G_e \tau_{\text{eff}}. \tag{2.128}$$

The approximation represented by Eq. (2.127) is invalid in some important cases such as the diffusion in the region of p-n junctions in semiconductors and in the surface layers of semiconductors and insulators.

We shall now consider a quantitative description of relaxation in ionic crystals irradiated with soft x rays (characterized by a photon energy of about 100 eV).[81] The absorption coefficient μ of such x rays in solids is of the order of 10^5 cm^{-1}, so that the primary excitations penetrate effectively a layer of a crystal which is about $\mu^{-1} \approx 100$ nm thick. Secondary fast electrons created by photons of this energy have a mean free path which in the case of ionization amounts to 1–2 nm (see Fig. 2.19). The number of newly created electron–hole pairs is described well by Eq. (2.102) with the value $E_{eh} \approx 15$ eV and it agrees with the estimate given by Eq. (2.111). We can therefore assume that directly after the end of electron relaxation in a crystal there are few secondary excitations per each absorbed photon and these are within regions of radii amounting to about 5 nm around the points of absorption of the primary photons. Since these radii are considerably less than the dimensions of the electron–phonon relaxation region, it follows that the size of the region of electron relaxation of a primary excitation is pointlike compared with the region of total relaxation. This makes it possible to write down the function representing secondary carrier generation:

$$G(x) = \left| \frac{dI}{dx} \right|_x E_{eh}^{-1} = \frac{I_0 \mu}{E_{eh}} \exp(-\mu x),$$

where x is the distance from the irradiated surface of a solid.

We shall consider the case when irradiation lasts for a time considerably longer than τ_{eff} and we shall write down the diffusion equation (2.125) in the quasisteady form:

$$D_{eff} \frac{d^2 N(x)}{dx^2} - \frac{N(x)}{\tau_{eff}} + \frac{I_0 \mu}{E_{eh}} \exp(-\mu x) = 0. \qquad (2.129)$$

An analogous equation (valid when $E_{eh} = \hbar\omega$) is fully applicable to the photoconductivity of semiconductors in the case when $\hbar\omega < E_g + E_t$.

We shall seek the solution in the form

$$N(x) = C_1 \exp\left(-\frac{x}{L_{eff}}\right) + C_2 \exp\left(\frac{x}{L_{eff}}\right) + \frac{I_0 \mu \tau_{eff}}{E_{eh}(1 - \mu^2 L_{eff}^2)} \exp(-\mu x). \qquad (2.130)$$

We can determine the constants C_1 and C_2 by formulating the boundary conditions, i.e., by specifying the situation on the boundaries of a crystal. We shall consider a semi-infinite crystal (assuming that its thickness is $d \gg L_{eff}$ and $d \gg \mu^{-1}$). The flux of carriers across the front (irradiated) surface of a crystal is

$$j_0 = -D_{eff} \frac{dN}{dx} \bigg|_{x=0} = N|_{x=0} s_0, \qquad (2.131)$$

where s_0 is the rate of loss of carriers on this surface [representing, for example, the losses due to electron emission (Sec. 2.6) or due to surface recombination]. It follows from the conditions of the problem that on the opposite surface we have

$$-D_{eff} \frac{dN}{dx} \bigg|_{x \to \infty} = 0, \qquad (2.132)$$

which leads directly to $C_2 = 0$. The simultaneous solution of Eqs. (2.130)–(2.132) is

$$N(x) = \frac{I_0 \tau_{eff} \mu}{E_{eh}(\mu^2 L_{eff}^2 - 1)} \left[\frac{s_0 + \mu D_{eff}}{s_0 + L_{eff} \tau_{eff}^{-1}} \exp\left(-\frac{x}{L_{eff}}\right) - \exp(-\mu x) \right]. \qquad (2.133)$$

Integration of Eq. (2.133) over the thickness of a crystal gives the following total quasisteady number of carriers in a crystal:

$$N_\infty = \int_0^\infty N(x) dx = \frac{I_0 \tau_{eff} \mu}{E_{eh}(\mu^2 L_{eff}^2 - 1)} \left(\frac{s_0 + \mu D_{eff}}{s_0 + L_{eff} \tau_{eff}^{-1}} L_{eff} - \mu^{-1} \right). \qquad (2.134)$$

If the rate of surface losses of carriers is high ($s_0 \gg \mu D_{eff}$, $s_0 \gg L_{eff}\tau_{eff}^{-1}$), the characteristics of the surface are eliminated from the above expression:

$$N_\infty = \frac{I_0 \tau_{eff}}{E_{eh}(1 + \mu L_{eff})}. \qquad (2.135)$$

If there is some method which can be used to find N_∞ or any quantity proportional to it, then the knowledge of μ allows us to use Eq. (2.135) to determine

L_{eff} quantitatively. In the case of many ionic crystals the values of L_{eff} found in this way are within the range 10–10^3 nm. The knowledge of L_{eff} and some data on the band structure of a crystal allows us to estimate the average values of other parameters representing the motion of hot carriers in ionic crystals in the course of their electron–phonon relaxation. Typical values of these parameters are $\tau_{eff} \approx 1$ ps, $l_{ev} \approx 10$ nm, $D_{eff} \approx 10^2$ cm^2/s (Ref. 81).

If the rate of surface lossses is low ($s_0 \ll \mu D_{eff}$, $s_0 \ll L_{eff}\tau_{eff}^{-1}$) or if the absorption coefficient is small ($\mu L_{eff} \ll 1$) the surface of a crystal does not affect the fate of carriers. In such cases Eq. (2.134) becomes

$$N_\infty = \frac{I_0 \tau_{eff}}{E_{eh}}, \tag{2.136}$$

which is identical with Eq. (2.128) if we bear in mind that $G_e = I_0 E_{eh}^{-1}$.

A comparison of Eqs. (2.135) and (2.136) allows us to find the number of carriers that leak away from the bulk of a crystal to its surface in the case when $s_0 \gg \mu D_{eff}$ and $s_0 \gg L_{eff}\tau_{eff}^{-1}$:

$$\Delta N_\infty = \frac{I_0 \tau_{eff}}{E_{eh}} \frac{\mu L_{eff}}{1 + \mu L_{eff}}. \tag{2.137}$$

Therefore, if irradiation of a crystal with photons occurs under conditions such that $\mu L_{eff} \gtrsim 1$ and $s_0 \gg \mu D_{eff}$, $s_0 \gg L_{eff}\tau_{eff}^{-1}$, then a large fraction of carriers generated by the radiation leaks away to the surface of a crystal and is lost to any processes in the interior. These carriers can be detected as a result of electron emission (see Sec. 2.6) or surface luminescence. On the other hand, if the above inequalities are not obeyed, then instead of Eq. (2.125), we can right from the beginning employ a much simpler equation (2.127).

Carriers generated in solids in the course of irradiation are manifested most directly by a change in the electronic conductivity of the investigated object. If the carrier density in a crystal in the absence of irradiation is N_e and the radiation-induced increase is ΔN_e, then the electrical conductivity can be described by

$$\xi_e = e(N_e + \Delta N_e)\mu_e, \tag{2.138}$$

where $\mu_e = e\tau_{er}/m_e$ is the carrier mobility and τ_{er} is the carrier relaxation time. Using $\tau_{er} = l_e/v_e$, $l_e^{-1} = l_{ev}^{-1} + l_{es}^{-1}$, $l_{es} = \sigma_{es}N_s$, and also the fact that the defect concentration is $N_s = N_{s0} + \Delta N_s$ (here, N_{s0} is the concentration of defects before irradiation and ΔN_s is the concentration of defects generated by irradiation—see Chap. 3), we find that Eq. (2.138) becomes

$$\xi_e = \frac{e^2 l_{ev}(N_e + \Delta N_e)}{m_e v_e(l_{ev}\sigma_{es}N_{s0} + l_{ev}\sigma_{es}\Delta N_s + 1)}. \tag{2.139}$$

It follows from Eq. (2.139) that irradiation can increase or decrease the value of ξ_e and that the direction and magnitude of the changes depend on the ratio of the increase ξ_e due to the newly generated carriers, on the one hand, and its suppression by the generated defects, on the other. The nature of such changes may be very different for different classes of solids.

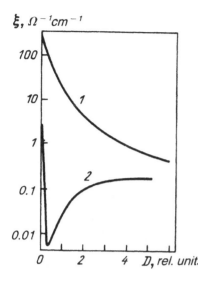

$\xi, \, \Omega^{-1} cm^{-1}$

FIG. 2.34. Changes in the electrical conductivity of InSb (1, n-type sample; 2, p-type sample) with the neutron radiation dose.[12]

Metals always contain a large number of free carriers so that we have $N_e \gg \Delta N_e$ and the change in ξ_e as a result of irradiation is determined mainly by the restrictions on the carrier mobility imposed by the newly created defects. Therefore, irradiation of metals practically always reduces their electrical conductivity. A typical magnitude of the effect is a change by 1–3 $\mu\Omega$ cm per 1% of defect sites.[73]

Interesting effects can also be observed in semiconductors. In typical cases the density of free carriers is 10^{15}–10^{18} cm^{-3} in the absence of irradiation. Radiation defects created in the lattice act as the centers of capture and recombination of electrons and holes, and they reduce the steady-state value of N_e and thus the value of ξ_e. A reduction in the mobility μ_e, due to the scattering by the new defects, also reduces the conductivity. Therefore, the resistivity of semiconductors usually rises as a result of irradiation.

There are, however, interesting exceptions. For example, the conductivity of a p-type semiconductor is frequently reduced first by irradiation and then rises (Fig. 2.34). This is due to the fact that at high concentrations of radiation defects the p-type conductivity may become n-type and this tends to a certain limiting value. This is known as the conduction inversion.

In the case of insulators and semiconductors with a wide band gap there is no electronic conductivity under equilibrium conditions ($N_e = 0$). It is manifested only during irradiation by ionizing radiations and it is due to newly generated carriers: it causes a serious deterioration of insulating properties of such materials during irradiation.

It follows from Eqs. (2.138) and (2.136) that for a given rate of carrier generation the electronic conductivity of wide-gap solids is governed by the factor $\mu_e \tau_{\text{eff}}$ which depends on the properties of a specific crystal. For example, in the case of irradiation with high-energy electrons, we find that typical values of $\mu_e \tau_{\text{eff}}$ for a covalent semiconductor (diamond) is 10^{-6} cm^2/V ($\tau_{\text{eff}} \approx 10^9$ s, $\mu_e \approx 10^3$

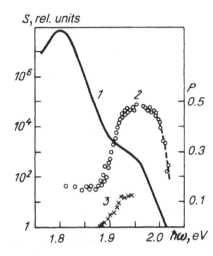

FIG. 2.35. Recombination luminescence spectrum (1) compared with the spectral dependences of the degree of circular (2) and linear (3) polarizations of the luminescence emitted by p-type $Al_{0.25}Ga_{0.75}As$ excited by polarized radiation of $\hbar\omega = 1.96$ eV energy.[24]

$cm^2\,V^{-1}\,s^{-1}$),[33] whereas for an insulator SiO_2 we have 10^{-11}–10^{-12} cm^2/V ($\tau_{eff} \approx 10^{-12}$ s, $\mu_e \approx 1$–10 $cm^2\,V^{-1}\,s^{-1}$).[94]

Carriers in wide-gap semiconductors and insulators can be created also after the end of irradiation via liberation of carriers localized in the course of irradiation. Such liberation may be achieved by heating a crystal [see Eq. (2.118)] and by photoionization of local centers. The conductivity due to such carriers is known, depending on the carrier liberation method, as the thermally stimulated or the photostimulated conductivity.

One of the most typical phenomena manifested by solids as a result of recombination of radiation-generated electrons and holes and also in the course of exciton annihilation is known as the recombination luminescence.

The most common type of the recombination luminescence is the radiation generated by the interband recombination of electrons and holes, frequently via an intermediate exciton state. The corresponding luminescence band is located in the region of the low-energy edge of the fundamental absorption region of a given substance. The profiles of the spectra reflect the quasiequilibrium energy distribution of the conduction electrons and valence holes or of free excitons. Figures 2.35 and 2.36 show the spectra of such luminescence for a semiconductor (p-type AlGaAs) and an insulator (NaI). In the former case we can see a low-energy band with a profile reflecting the Maxwellian distribution of thermalized electrons and a high-energy band polarized in accordance with the polarization of the exciting light and due to hot electrons which have not yet lost the initial spin orientation. In the latter case the spectrum has a band at 5.6 eV, which is due to free excitons [Fig. 2.36(a)]. It consists of a zero-phonon free-exciton line and its phonon replicas [Fig. 2.36(b)].

Luminescence of this type had been observed also for metals. For example, irradiation of copper with 500-eV electrons showed that the integral yield at wavelengths 240–700 nm is 10^{-8} photons per electron per steradian per electron volt.[31]

If the irradiation is intense, the spontaneous recombination luminescence may

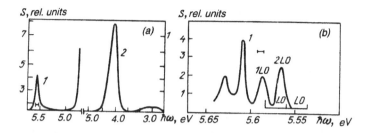

FIG. 2.36. Luminescence spectra of NaI irradiated with electrons (*a*, Ref. 35) and with $\hbar\omega = 5.75$ eV photons (*b*, Ref. 47): (1) free-exciton luminescence; (2) self-localized exciton luminescence.

become stimulated. This is manifested by a strong narrowing of the emission band demonstrated in Fig. 2.37. Such stimulated radiation had been observed experimentally for a number of semiconductors [GaAs at 835 nm (Ref. 8), CdSe at 695 nm (Ref. 43), and GaPAs at 704 nm (Ref. 7)] irradiated with 100-keV electrons using current densities in excess of 1–2 A/cm². The necessary quasisteady carrier density amounts to 10^{15}–10^{17} cm⁻³.

In ionic crystals in which holes and excitons become self-localized, a strong luminescence appears as a result of recombination of electrons with the V_k centers (self-localized holes). All alkali halide crystals have a band in the near-ultraviolet or visible parts of the spectrum and this band is due to the ${}^3\Sigma_u^+ \to {}^1\Sigma_g^+$ transition in the X_2^{2-} quasimolecule (this is known as the π luminescence and in the case of NaI it occurs at 2.5 eV). Some crystals reveal also a band at a higher energy corresponding to the ${}^1\Sigma_u^+ \to {}^1\Sigma_g^+$ transition (this is known as the σ luminescence and it occurs at 4.1 eV), as illustrated in Fig. 2.36(a).

When impurities are present in crystals, an impurity recombination luminescence may be observed and it may compete with the intrinsic luminescence. The impurity luminescence spectrum is governed by transitions in impurity atoms or ions.

The result of radiative tunnel transitions is a tunnel luminescence with the photon energy governed by Eq. (2.119). A characteristic manifestation of the tunnel luminescence is a weak temperature dependence of the intensity (thermal liberation of carriers is not needed) and a shift of the luminescence band maximum with time (which occurs because close pairs, i.e., those with a short internal distance, recombine faster, so that the effective value of R increases with time).

A surface recombination luminescence is the result of radiative recombination of carriers on crystal surfaces. It increases in intensity on reduction in the depth of generation of carriers. In the case of photoexcitation, the excitation spectrum of this luminescence is proportional to ΔN_∞ of Eq. (2.137) and antibatic to the bulk recombination luminescence spectrum, where the intensity is proportional to N_∞ of Eq. (2.135) (an increase in the absorption coefficient of the exciting radiation increases also the probability that carriers migrate from the bulk to the surface).[81]

The recombination luminescence of solids excited by ionizing radiations has a wide range of practical applications. Luminescent screens for the visualization of images in x-ray systems have been known for a long time. The main materials in

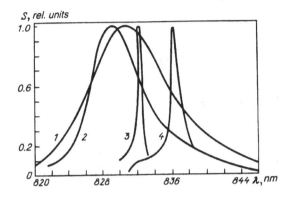

FIG. 2.37. Luminescence spectrum of GaAs irradiated with 50-keV electron beams of different current densities (A/cm²): (1) 0.5; (2) 1.1; (3) 2.2; (4) 3.5. Data taken from Ref. 8.

such screens are ZnS:Ag, CaWO$_4$, Ba$_3$(PO$_4$)$_2$:Eu, CsI:Na, Y$_2$O$_2$S:Yb, and other substances which are mainly insulators.[20] Scintillation counters, in which a working substance is a single crystal of NaI:Tl, CsI:Tl, anthracene, stilbene, etc., as well as some liquid or plastic organic substances, play an important role in detection and measurements of ionizing radiations.[15]

The recombination luminescence appears also as a result of recombination of carriers liberated in the course of annealing of irradiated crystals with localized carriers of the opposite sign. This is known as thermoluminescence. The temperature dependence of its intensity is a nonmonotonic curve with maxima at temperatures corresponding to the fast release of carriers. The thermoluminescence spectrum is governed by optical transitions at the recombination centers. Similarly, the recombination luminescence may appear also as a result of optical liberation of localized carriers. The ability of many substances to retain for a long time a stored light sum and then liberate it in the form of the thermally stimulated or photostimulated luminescence is used in dosimetry of ionizing radiations.[55]

2.6. Electron emission from solids

It follows from Secs. 2.4 and 2.5 that crystals bombarded with ionizing radiations always have a certain number of free electrons. Under certain conditions some of these electrons may leave a crystal. This is known as electron emission and it underlies many practical applications of solids; moreover, in principle, this phenomenon can provide extensive information on physical properties of the emitting crystals.

In the first approximation, the process of emission of an electron from a crystal can be regarded as a sequence of three elementary stages: creation of a free electron, its approach to the surface of a crystal, and escape from the crystal into vacuum (Fig. 2.38). If the rate of generation of free carriers is G_e and the probabilities of the two successive stages are P_s and P_v, respectively, the electron emission rate can be described by

$$S_e = G_e P_s P_v. \tag{2.140}$$

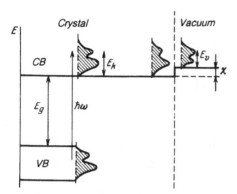

FIG. 2.38. Three-stage model of the photoelectric effect: VB is the valence band; CB is the conduction band; χ is the work function; E_k and E_v are the kinetic energies of an electron in a crystal and in vacuum, respectively.

This is known as the three-stage model of electron emission. The first stage of this process (creation of a free electron) is due to the primary or secondary interactions of ionizing radiation with a solid discussed in the preceding sections.

We shall now consider the process of emission of an electron from the surface of a crystal into vacuum. In the case of the majority of solids an electron must overcome a certain potential barrier χ known as the work function, which is numerically equal to the electron affinity of a crystal. The law of conservation of energy for this process is

$$E_v = E_k - \chi, \tag{2.141}$$

where E_v and E_k are the values of the kinetic energy of an electron in vacuum and in a crystal near its surface.

The consequences of the law of conservation of the momentum can be considered conveniently by expanding the momentum into its components, one of which ($\mathbf{p}_\parallel$) is parallel to the surface of a crystal, whereas the other ($\mathbf{p}_\perp$) is perpendicular to the surface (Fig. 2.39). We then obtain

$$\mathbf{p} = \mathbf{p}_\perp + \mathbf{p}_\parallel,$$

where

$$p_{k\perp} = p_k \cos\beta, \quad p_{k\parallel} = p_k \sin\beta, \tag{2.142a}$$

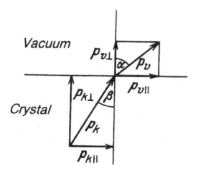

FIG. 2.39. Change in the momentum of an electron on escape from a crystal in vacuum.

$$p_{v\perp}=p_v \cos \alpha, \quad p_{v\parallel}=p_v \sin \alpha. \tag{2.142b}$$

Here, β and α are the angles at which an electron approaches the surface and is emitted from a crystal, whereas the indices k and v apply to the crystal and vacuum, respectively.

The second law describing the emission of an electron from a crystal is the law of conservation of $\mathbf{p}_\parallel$, i.e.,

$$\mathbf{p}_{v\parallel}=\mathbf{p}_{k\parallel}+\mathbf{g}, \tag{2.143}$$

where $\mathbf{g}$ is the reciprocal lattice vector. This law is a consequence of the translational symmetry of the surface of a crystal and it therefore postulates a high degree of order of the surface.

We shall consider the case when the emission of an electron from a crystal does not result in the transfer of the momentum to the crystal. It then follows from Eqs. (2.141) and (2.143) that the selection rule for the electron emission is

$$\hbar^2 q_{k\perp}^2/2m \geqslant \chi. \tag{2.144}$$

This means that an electron may be emitted from a crystal if it has a certain minimum value of the quasimomentum directed toward the surface:

$$q_{k\perp\text{min}}=(2m\chi/\hbar^2)^{1/2}.$$

It now follows from Eqs. (2.141)–(2.143) that

$$E_k^{1/2} \sin \beta = (E_k-\chi)^{1/2} \sin \alpha, \tag{2.145}$$

which means that $\alpha > \beta$. Consequently, there is a certain maximum value $\beta = \beta_m$ corresponding to $\sin \alpha = 1$ for which an electron may be emitted from a crystal; β_m is given by

$$\sin \beta_m=(1-\chi/E_k)^{1/2}. \tag{2.146}$$

It therefore follows that in the case of a transition of an electron from one medium to another, which differs from the former by the electron affinity, there are effects analogous to the refraction and total internal reflection in optics.

Equation (2.144) predicts a multistage nature of the probability of emission of an electron from a crystal into vacuum, depending on the value of $q_{k\perp}$:

$$P_v=\begin{cases}1 & \text{if} & \hbar^2 q_{k\perp}^2/2m \geqslant \chi, \\ 0 & \text{if} & \hbar^2 q_{k\perp}^2/2m < \chi.\end{cases} \tag{2.147}$$

A more rigorous analysis of the emission of an electron from a crystal must allow for above-barrier reflection and tunneling across the surface barrier, the effect of which "round out" somewhat the Heaviside function of Eq. (2.147).

The nature of the second stage of electron emission, which is an approach of an electron from the bulk of a crystal to its surface, depends strongly on the distance x from the point of creation of a free electron to the surface. The probability that an electron with a mean free path l_e undergoes k collisions in the distance x is given by the Poisson distribution:

$$P_k(x) = \frac{1}{k!} \left(\frac{x}{l_e} \right)^k \exp\left(-\frac{x}{l_e} \right). \tag{2.148}$$

If we assume an isotropic distribution of the momenta of free electrons at the moment of their creation, we find that the probability of arrival of an electron at the surface without collisions is

$$P_s = \frac{1}{2} \exp\left(-\frac{x}{l_e'} \right), \tag{2.149}$$

where $l_e' = l_e \cos \beta$ and the factor $\frac{1}{2}$ represents the probability that an electron is directed toward the surface.

It therefore follows that electrons liberated by radiation in a surface layer of thickness comparable with l_e have a high probability of reaching the crystal surface retaining their energy and momentum which they acquire at the moment of creation.

If $x \gg l_e$, the majority of electrons reaches the surface after a multitude of collisions and they lose completely the information on the initial values of the energy and momentum. In this case the probability of arrival of an electron at the surface is again described by an exponential law:

$$P_s' = A \exp(-x/L_e), \tag{2.150}$$

where A is a constant value; L_e is the escape depth of electrons such that $L_e \gg l_e$. If $l_e < x < L_e$, an electron experiences a small number of collisions before it emerges from a crystal. In this case the value of P_s is not an exponential function of x, but a more complex one.

In quantitative estimates we must bear in mind that if, depending on the actual situation (substance, value of E_k), the mean free path l_e is governed either by the electron–electron or the electron–phonon scattering, and we will generally have $l_e^{-1} = l_{ee}^{-1} + l_{ev}^{-1}$ (Secs. 2.3 and 2.5).

One of the most interesting types of electron emission is the primary photoelectron emission, sometimes simply called photoemission.

We shall consider a layer of a crystal of thickness dx at a distance x from the surface along the normal to the latter. When a flux of photons S_0 is incident normally on the crystal and each of these photons creates one free electron, the rate of generation of primary photoelectrons in the surface layer is given by [see Eq. (2.48)]:

$$dG_e = S_0 \mu \exp(-\mu x) dx. \tag{2.151}$$

The rate of primary electron emission S_{ep} is obtained from Eqs. (2.140), (2.149), and (2.151):

$$S_{ep} = \int_0^\infty dG_e P_s P_v = \frac{1}{2} S_0 \mu P_v \int_0^\infty \exp[-x(\mu + l_e^{-1})] dx = \frac{S_0 P_v \mu l_e}{2(1 + \mu l_e)}. \tag{2.152}$$

When solids are irradiated with ultraviolet photons or with soft x rays, we usually have $\mu l_e \ll 1$ ($\mu \approx 10^5 – 10^6 \text{ cm}^{-1}$, $l_e \approx 10^{-7} \text{ cm}$). It then follows from Eqs. (2.152), (2.46), and (2.53) that

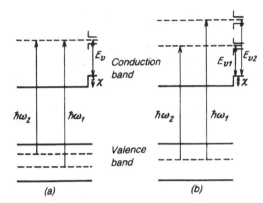

FIG. 2.40. Investigation of the energy band structure of a crystal by photoelectron spectroscopy methods. Electrons with the kinetic energy E_v are detected; if $E_v = \text{const}$ and $\hbar\omega$ is varied, the valence band is probed (a), but if E_v and $\hbar\omega$ vary so that $\hbar\omega_1 - \hbar\omega_2 = E_{v1} - E_{v2}$, the conduction band is probed (b).

$$\frac{S_{ep}}{S_0} = \frac{1}{2} P_v \mu l_e = N_0 P_v l_e \frac{2\pi^2 e^2}{m^2 c \omega} |M'_{f0}|^2 \rho_q(E_f). \qquad (2.153)$$

If we bear in mind that the factor $\rho_q(E_f)$ ensures that the law of conservation of energy is obeyed in the absorption of a photon and the probability P_v differs from zero only if $E_f \geqslant \chi$, we find that Eq. (2.153) predicts the characteristic features of photoelectron emission which are the proportionality to the incident photon flux and to the square of the amplitude of the electric field, and the existence of a threshold energy $\hbar\omega_t = E_g + \chi$ (see Fig. 2.38). This conclusion is of major methodological importance because it shows that photoelectron emission can be explained without invoking quantization of radiation (the need for such quantization is frequently assumed), but simply by quantizing the energy states of matter, as is done in quantum mechanics.

In the case of moderately high incident photon energies, when the law of conservation of the quasimomentum of an electron of Eq. (2.52) is obeyed, an investigation of the dependence $S_{ep}(\hbar\omega)$ provides extensive information on the band structure of a substance. If only electrons with a specific kinetic energy E_v are detected, the rate of electron detection is $N_e(\hbar\omega, E_v) = S_{ep}(\hbar\omega)\delta(E_f - E_v)$, where $\delta(E_f - E_v)$ is used to select the final states with the kinetic energy E_v. If E_v is fixed, it follows from the condition (2.52) that the $N_e(\hbar\omega; E_v)$ spectrum is proportional to the density of such states in the valence band $\rho_h(E)$ from which direct transitions characterized by $E_f = E_v$ are possible. If E_v and $\hbar\omega$ change simultaneously and the difference $\hbar\omega - E_v$ is constant, the resultant dependence $N_e(\hbar\omega; \hbar\omega - E_v)$ reflects the density of states in the conduction band $\rho_e(E)$ to which direct transitions are possible from the valence-band states of energy $E_0 = E_f - \hbar\omega$ (Fig. 2.40).

Even more detailed information on the energy band structure of crystals can be obtained if, in addition to the electron energy, we determine also their angular distribution. In this case it follows from Eqs. (2.142) and (2.143) that

$$q_{k\parallel} = \left(\frac{2mE_v}{\hbar^2}\right)^{1/2} \sin \alpha. \qquad (2.154)$$

For each energy E_v and each angle α only those electrons appear in the spectrum which have an initial energy $E_0 = E_v - \hbar\omega$ and which correspond to a reciprocal

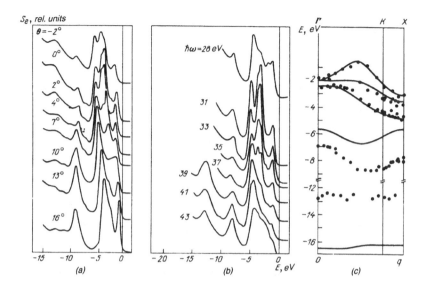

FIG. 2.41. Spectra of photoelectrons emitted from the (100) surface of PbSe as a function of the angle of emergence for the constant incident photon energy (35 eV) (a) and as a function of the energy of the incident photons at a constant angle (normal to the surface) of emission (b). The energy band structure parallel to the (100) plane plotted on the basis of these spectra is shown in (c) (Ref. 104).

lattice line defined by Eq. (2.154). Variation of the value of α and/or $\hbar\omega$ allows us to find the dependence $E_0(q_{k\|})$, i.e., the dispersion of the valence band parallel to the investigated surface (Fig. 2.41). The dispersion of the electron states perpendicular to the surface remains indeterminate and this is a consequence of breakdown of the translational symmetry along this direction because of the presence of the surface. In crystals with a layer structure there is no dispersion perpendicular to the layers and the dependence $E_0(q_{k\|})$ determines completely the energy band structure. In the case of a three-dimensional structure the experimental data described above can be supplemented by theoretical ideas or calculations.

At high incident photon energies, when the rule (2.52) loses its force, the distribution of the photoelectron energies E_v reflects the combined density of the initial and final states without resolution in respect to the quasimomentum. Since at these energies the density of the final states is a continuous function of the energy, the distribution reproduces the distribution of electrons in the ground state of a crystal (or any ionizable object) in respect to the binding energy E_{nl}:

$$E_{nl}=\hbar\omega-E_v-\chi. \tag{2.155}$$

This circumstance is the basis of the method of x-ray electron spectroscopy of matter. This method is currently being used to determine the values of E_{nl} for various electron shells of free or bound atoms to within 0.1–0.5 eV (Fig. 2.42). Such precision is sufficient to detect changes in E_{nl} in the case of the inner electron shells as a result of a change in the state of the outer electrons because of the participation of the latter in chemical binding. This is the reason for the widely used

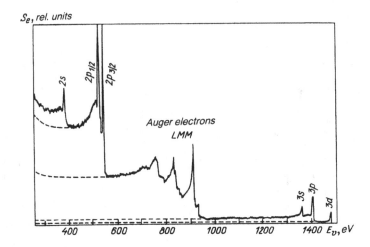

FIG. 2.42. X-ray electron spectrum of copper irradiated with K x rays from Al ($\hbar\omega = 1.49$ keV). The ionizable electron shells are shown and an increase in the number of the scattered electrons in the direction of lower energies is demonstrated.[124]

name for the method: electron spectroscopy for chemical analysis (ESCA). In the case of solids this method can be used to determine also the density of states function in the valence band (see Fig. 2.22).

The energy resolution of this electron spectroscopy method can be improved if electrons are generated by monochromatic atomic emission lines, for example, the resonance line of helium at $\hbar\omega = 21.22$ eV. This makes it possible to investigate the vibronic structure of the valence states of small molecules. Figure 2.43 shows a photoelectron spectrum of the N_2 molecule and a system of potential energy curves accounting for the origin of this spectrum. We can see transitions to various electronic states of the molecule as a function of the quantum number of the electron being removed ($3\sigma_g$, $1\pi_u$, or $2\sigma_u$) and the vibrational levels of these states. The energies of the corresponding lines are given by

$$E_v = \hbar\omega - E_0 - (E_{vib}^+ - E_{vib}^0)$$

(where E_{vib}^+ and E_{vib}^0 are the vibrational energies of a molecule before and after an electron transition), whereas the intensities are governed by the Franck–Condon factors (see Sec. 1.2). A series of vibrational lines has a number of terms which increases with the difference between the equilibrium internuclear states in the normal and ionized molecules.

The emission of electrons generated in solids in the course of the electron–electron stage of relaxation primary excitations is known as the secondary electron emission. As a rule, secondary electrons approach the surface of a crystal after many electron–phonon (and also electron–defect) collisions, and they do not carry any information on the event responsible for the creation of the primary excitation.

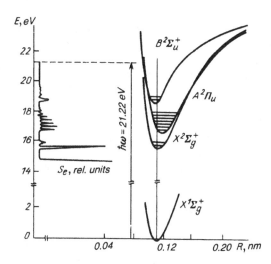

FIG. 2.43. Potential energy curves of different states and the photo-electron spectrum (formed as a result of the absorption of photons of $\hbar\omega = 21.22$ eV energy) of the N_2 molecule.[62]

Physical information obtainable from the secondary electron emission spectra depends on the method used to record the electrons emitted from a crystal. We can distinguish at least three different recording regimes:[79]

(1) all the emitted electrons are detected independently of one another, which can be achieved in the case of a high (of the order of 1 ps) time resolution in electron counting or by measurements of the emission current; the corresponding emission efficiency is known as the current quantum efficiency κ_c;

(2) all the electrons emitted after absorption of one photon or one particle by a crystal are recorded as one emission event; this can be done using a moderate time resolution in electron counting and in determination of the incident radiation flux (for example, 10^6 s^{-1}); the emission efficiency is known as the pulsed quantum efficiency κ_p;

(3) only those emission events are recorded which represent exactly n electrons (where $1 \leqslant n \leqslant n_0$, where n_0 is the number of electrons created in a crystal by the absorption of one photon or electron); this can be achieved using good proportional electron counters; the corresponding emission efficiency is called the n-electron quantum efficiency κ_n.

If the probability that a conduction electron migrates a distance x from the point of its creation is assumed, following Eq. (2.148), to be $\exp(-x/ \to x/L_m)$, where L_m is the average migration distance, then postulating that this electron loses the whole of its kinetic energy quasicontinuously until thermalization and then stops (it is captured or recombines), the probability that an electron with an initial kinetic energy E_0 has a kinetic energy E_x after migrating a distance x is $\exp\{(-x/L_m)[E_0/(E_0-E_x)]\}$. The probability of recording an emitted electron created with an energy E_0 at a depth of x from the front surface of a crystal is now given by

$$P_1 = pA \exp\left(-\frac{x}{L_m}\frac{E_0}{E_0-\chi}\right) = pA \exp\left(-\frac{x}{L_e}\right), \qquad (2.156)$$

where A is the probability that after migration over a distance x, equal to the depth of generation of an electron, the latter is on the surface; p is the probability of detection of the emitted electron; $L_e = L_m(1 - \chi/E_0)$.

We can use Eq. (2.156) to find readily the probability of detection of at least one electron after n_0 electrons are created in a crystal at a depth x:

$$B = 1 - (1 - P_1)^{n_0} = 1 - \left[1 - pA \exp\left(-\frac{x}{L_e}\right)\right]^{n_0} \tag{2.157}$$

[the probability of failure to detect one electron is $(1 - P_1)$ and the probability of failure to detect n_0 electrons is $(1 - P_1)^{n_0}$], whereas the probability of detecting exactly n of these electrons is

$$B_n = C_{n_0}^n \left[pA \exp\left(-\frac{x}{L_e}\right)\right]^n \left[1 - pA \exp\left(-\frac{x}{L_e}\right)\right]^{n_0 - n}. \tag{2.158}$$

Now, knowing the distribution of electrons with respect to x or, in other words, the generation function $G_e(x)$, we can find the quantum efficiencies of emission mentioned above. This is simplest to carry out in the case of photon irradiation. Then, using Eq. (2.151) and introducing the concept of the grazing incidence angle ϑ of radiation (i.e., of the angle between the surface of the crystal and the direction of incidence of the radiation), we find that

$$\kappa_c = \frac{n_0}{S_0} \int_0^\infty P_1 dG_e(x) = \frac{n_0 \mu p A}{\cos \vartheta} \int_0^\infty \exp\left(-\frac{\mu x}{\cos \vartheta} + \frac{x}{L_e}\right) dx = \frac{n_0 p A L_e' \mu}{1 + \mu L_e'}, \tag{2.159}$$

$$\kappa_p = \frac{1}{S_0} \int_0^\infty B dG_e(x) = \sum_{n=1}^{n_0} (-1)^{n+1} C_{n_0}^n (pA)^n \frac{\mu L_e'}{n + \mu L_e'}, \tag{2.160}$$

$$\kappa_n = \frac{1}{S_0} \int_0^\infty B_n dG_e(x) = \frac{\mu L_e' C_{n_0}^n (pA)^n (1 - pA)^{n_0 - n}}{n + \mu L_e'} + \frac{n + 1}{n + \mu L_e'} \kappa_{n+1}, \tag{2.161}$$

where $L_e' = L_e \cos^{-1} \vartheta$. The expression (2.160) simplifies greatly if $pA < n_0^{-1}$:

$$\kappa_p (pA < n_0^{-1}) = \frac{n_0 p A \mu L_e'}{1 + \mu L_e'} = \kappa_c.$$

It follows from Eqs. (2.159) and (2.160) that determination of the dependences $\kappa_c(\vartheta)$ and $\kappa_p(\vartheta)$ for a given value of μ or of $\kappa_c(\mu)$ and $\kappa_p(\mu)$ for a given value of ϑ makes it possible to determine quantitatively the value of L_e and the factor $n_0 pA$. It is clear from Eq. (2.161) that the most informative is the determination of $\kappa_n(\vartheta)$ or $\kappa_n(\vartheta)$, because it allows us to determine separately L_e, pA, and n_0.

This methodology has been used particularly in studies of alkali halide crystals irradiated with ultrasoft x rays. A typical value of L_e is found to be 30–100 nm and $E_{eh} = \hbar\omega/n_0$ is 12–17 eV, in good agreement with the estimate provided by Eq. (2.111) (Ref. 79).

In the case of irradiation with electrons the function $G_e(x)$ is given by Eq. (2.89) and the problem is more complex than in the case of photon irradiation. In simple

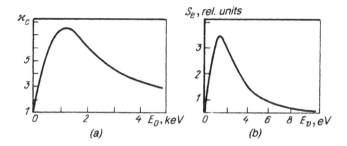

FIG. 2.44. Dependence of the number of secondary electrons on the energy of the incident electrons (a) and the spectrum of the secondary electrons (b) emitted from an MgO single crystal.[97,134]

estimates we can assume that if $0 \leqslant x \leqslant R$ [where R is the range of a primary electron given by Eq. (2.90)] the function $G_e(x)$ is independent of x and has the value G_0. Then, instead of Eq. (2.159) we find that κ_c for the normal incidence of primary electrons is described by

$$\kappa_c = \frac{pAG_0}{S_0} \int_0^R \exp\left(-\frac{x}{L_e}\right) dx = \frac{pAG_0 L_e}{S_0}\left[1 - \exp\left(-\frac{R}{L_e}\right)\right].$$

At low values of E_0 and R (when $R \ll L_e$) the efficiency $\kappa_c = pAG_0RS_0^{-1}$ is a slowly rising function of E_0 (R increases on increase in E_0 faster than G_0 decreases). At high values of E_0 the quantum efficiency $\kappa_c \approx pAG_0L_eS_0^{-1}$ varies inversely with E_0 (G_0 decreases on increase in E_0). For many substances the efficiency κ_c has its maximum (of about 10) at $E_0 \approx 2$ keV. Figure 2.44 shows, by way of example, the dependence of κ_c on E_0 and the energy distribution of electrons emitted from MgO.

If in the course of irradiation electrons in a solid become localized at trapping levels, electron emission may be stimulated even after the end of irradiation as a result of thermal or optical liberation of the localized electrons. This is known as the thermally stimulated or photostimulated electron emission. Such emission provides one of the most sensitive methods for the detection of small numbers of localized electrons in crystals.[98]

Chapter 3

Nuclear subsystem

3.1. Collisions of particles with nuclei

The primary process of the interaction of particles or photons with the nuclear subsystem in matter, which alters the state of this subsystem, is the displacement of one of the atoms from its initial position to a new nonequilibrium position. At moderately high values of the kinetic energy of the incident particles this process can be regarded as the result of a classical elastic pair collision of an incident particle with one of the atoms in matter (Sec. 1.1).

The condition for the occurrence of this process reduces to the requirement that the kinetic energy T transferred to an atom should exceed a certain threshold displacement energy E_d. The latter energy is, generally speaking, a function of the velocity of displacement, i.e., it is a function of T; moreover, in crystalline solids, it is also a function of the direction of displacement. The minimum value of E_d, corresponding to a slow displacement of an atom such that its environment can follow adiabatically the traveling atom, should be close to the cohesive energy of matter per one atom, i.e., it should be of the order of several electron volts. However, even at kinetic energies of several electron volts the time taken by an atom to travel one interatomic distance is shorter than the vibration period of atoms $\omega_0^{-1} \approx 10^{-13}$ s, so that the environment of the atom being displaced does not have a chance to relax and E_d can exceed considerably the minimum value just stated. Table 3.1 gives the measured values of E_d for a number of crystalline solids.

We shall consider the laws of conservation of energy and momentum in the case of an elastic pair collision when a particle of mass M_1 traveling at the velocity $\mathbf{v}_0$ interacts with a particle of mass M at rest, and after the interaction the incident particle has a velocity $\mathbf{v}_f$ and another particle acquires a velocity $\mathbf{v}$. We then have

$$M_1\mathbf{v}_0 = M_1\mathbf{v}_f + M\mathbf{v}, \quad M_1 v_0^2 = M_1 v_f^2 + M v^2.$$

A series of transformations gives

$$\frac{Mv^2}{2} = T = E_0 \frac{4M_1M}{(M_1+M)^2} \cos^2\varphi = T_m \cos^2\varphi = T_m \sin^2\frac{\theta}{2}, \quad (3.1)$$

where θ is the angle between $\mathbf{v}_0$ and $\mathbf{v}_f$; φ is the angle between $\mathbf{v}_0$ and $\mathbf{v}$; $E_0 = \frac{1}{2}M_1 v_0^2$; T_m is the maximum possible transferred energy:

TABLE 3.1. Threshold energy E_d necessary to displace an atom from a site to an interstice and the corresponding kinetic energies E_0^{min} of electrons and neutrons (protons) needed for such displacement in some crystals.[74]

Crystal	Displaced atom	E_d, eV	E_0^{min}, keV	
			e	$n(p)$
Ge	Ge	12–20	320–530	0.17–0.29
Si	Si	11–22	115–330	0.18
Diamond	C	80	530	0.29
InSb	In	6	250	0.14
	Sb	8	360	0.20
GaAs	Ga	9.0	230	0.13
	As	9.4	260	0.14
ZnS	Zn	10	240	0.13
	S	15–20	200	0.11
CdTe	Cd	5.6	235	0.13
	Te	7.8	340	0.18
MgO	O	60	320	0.17
BeO	O	76	400	0.22

$$T_m = E_0 \frac{4M_1 M}{(M_1 + M)^2} \approx 4E_0 \frac{M_1}{M} . \tag{3.2}$$

The last part of the above equality applies when $M \gg M_1$.

If a particle M is an atom in a solid with a mass much greater than M_1, then the minimum value of E_0 needed to displace this atom from its equilibrium position is governed by the condition $T_m = E_d$ and it amounts to

$$E_0^{min} = E_d \frac{(M_1 + M)^2}{4M_1 M} \approx \frac{E_d M}{4M_1} . \tag{3.3}$$

The values of E_0^{min} for a number of substances and various particles are also included in Table 3.1.

Figure 3.1 demonstrates the dependence of the efficiency of formation of defects in Ge on the energy of the incident electrons in the region of E_0^{min}.

If the incident particle has an electric charge, a collision can be regarded as a Coulomb collision of a particle with the nucleus of an atom and its cross section can be described by the Rutherford formula (1.17):

$$d\sigma = \left(\frac{Z_1 Z_2 e^2}{2\mu v_0^2}\right)^2 \sin^{-4} \frac{\theta}{2} d\Omega = \pi \left(\frac{Z_1 Z_2 e^2}{\mu v_0^2}\right)^2 \frac{T_m dT}{T^2} . \tag{3.4}$$

We shall assume that an atom is displaced always when $T \geqslant E_d$ and never if $T < E_d$. Then, the total cross section of the displacement process is

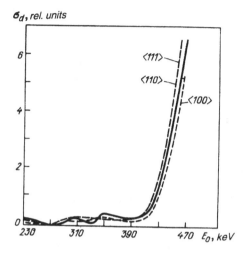

FIG. 3.1. Dependence of the cross section σ_d of the formation of defects in Ge on E_0 in the case of electron irradiation along the principal crystallographic axes.[68]

$$\sigma_d = \int_{E_d}^{T_m} d\sigma = \pi \left(\frac{Z_1 Z_2 e^2}{\mu v_0^2} \right)^2 \left(\frac{T_m}{E_d} - 1 \right) \approx \pi a_B^2 (Z_1 Z_2)^2 \frac{M_1}{M} \frac{E_B^2}{E_d E_0}.$$

(3.5)

The last equality applies when $T_m \gg E_d$ and $M \gg M_1$.

It is clear from Eq. (3.5) that, in particular, the average radius of the interaction for collisions resulting in the displacement of atoms is comparable with a_B and, consequently, is much less than the interatomic distance in condensed matter under normal conditions. This justifies the use of the model of pair collisions when discussing radiation damage to condensed matter.

The average energy T_d transferred in collisions causing displacements is

$$T_d = \int_{E_d}^{T_m} T d\sigma \bigg/ \int_{E_d}^{T_m} d\sigma = \frac{E_d T_m}{T_m - E_d} \ln \frac{T_m}{E_d} \approx E_d \ln \frac{T_m}{E_d}. \qquad (3.6)$$

If the incident particles have relativistic velocities, the above relationships are modified. For example, instead of Eq. (3.2), we now have

$$T_{\text{rel}} = E_0 \frac{2(E_0 + 2M_1 c^2)}{M c^2} \sin^2 \frac{\theta}{2} = T \left(1 + \frac{E_0}{2M_1 c^2} \right). \qquad (3.7)$$

The above expressions for the collision cross sections are invalid in the case of neutrons because they do not carry any electric charge. If the interaction of a neutron with a nucleus does not cause a nuclear reaction, then in the first approximation we can assume that the interaction involves isotropic scattering by a hard sphere of radius R:

$$\sigma_n = 4\pi R^2 \approx 4\pi A^{2/3} (1.5 \times 10^{-13} \text{ cm})^2, \qquad (3.8)$$

where A is the atomic mass of the scattering atom. In this case all the values of the recoil energy between 0 and T_m are equally probable and $T_d = \frac{1}{2} T_m$.

TABLE 3.2. Typical characteristics of the displacement of atoms in a substance with a moderately high atomic number ($M = 50$, $E_d = 10$ eV) irradiated with different particles.[51]

Incident particle and its energy	T_d, eV	T_m, keV	Incident particle and its energy	T_d, eV	T_m, keV
Proton, 1 MeV	150	80	Electron, 1 MeV	80	4
Ion ($M_1 \approx M$), 50 keV	110	50	Neutron, 2 MeV	8×10^4	160

Typical values of T_d and T_m in a substance with moderately large atomic number ($M = 50$, $E_d = 10$ eV) are listed in Table 3.2 for various particles.

Bearing this point in mind, we find that the rate of the radiation-induced primary displacements is

$$G_d = \sigma_d N_0 S = \sigma_d N_0 N v_0, \qquad (3.9)$$

where N_0 is the concentration of the target atoms, S is the flux of the incident particles, N is their concentration, and v_0 is their velocity.

It is shown in Sec. 2.3 that the linear energy losses experienced by charged particles with a charge $Z_1 e$ due to ionization of matter amount to [see Eqs. (2.89) and (2.93)]

$$-\left(\frac{dE}{dx}\right)_{\text{ion}} = \frac{2\pi e^4 Z_1^2 N_0 Z_2}{mv^2} \ln \frac{2mv^2}{I_{\text{av}}}. \qquad (3.10)$$

Using Eqs. (3.5) and (3.6), we can calculate the linear energy losses experienced by charged particles due to the displacement of atoms in a target:

$$-\left(\frac{dE}{dx}\right)_{\text{disp}} = N_0 T_d \sigma_d = \frac{2\pi e^4 Z_1^2 Z_2^2 N_0}{Mv^2} \ln \frac{T_m}{E_d}. \qquad (3.11)$$

It follows from Eqs. (3.10) and (3.11) that at a given velocity of the incident particles we have

$$\frac{(dE/dx)_{\text{ion}}}{(dE/dx)_{\text{disp}}} \approx \frac{M}{mZ_2} \frac{\ln(2mv^2/I_{\text{av}})}{\ln(T_m/E_d)} \approx 10^3. \qquad (3.12)$$

Therefore, the passage of charged particles through matter results in expenditure of the bulk of the energy of the particles in the ionization process.

However, we must bear in mind that the ionization of a substance, i.e., the inelastic collisions with electrons accompanied by the transfer of a minimum energy of the order of E_g can occur only in the case of particles whose energy exceeds a certain threshold value amounting to about $E_g M_1 / 4m$ [see Eq. (3.3)] and which in the case of protons is, for example, 10^3–10^4 eV. The particles, whose kinetic energy is between E_0^{min} and this high value expend their energy mainly by displacements of the atoms in a target substance.

Collisions of particles with the atomic nuclei may result in nuclear reactions, i.e., these collisions may be inelastic. The cross section of such collisions is governed by

TABLE 3.3. Characteristics of some typical nuclear reactions caused by thermal neutrons (ΔE is the released energy).[51]

Reaction	ΔE, MeV	σ, 10^{-14} cm^2	Reaction	ΔE, MeV	σ, 10^{-14} cm^2
$Li^6(n, \alpha)H^3$	4.7	950	$Mg^{25}(n, \alpha)Ne^{22}$	0.4	0.27
$B^{10}(n, \alpha)Li^7$	2.7	3990	$U^{235}(n, \text{fission})$	200	582

the range of the nuclear forces ($\approx 10^{-13}$ cm) and typically amounts to 1–100 mb (1 barn, abbreviated to b, is 10^{-24} cm^2).

In the case of charged particles the elastic interaction cross section given by Eq. (3.5) at moderately high values of E_0 is much greater than the inelastic interaction cross section. However, since it is inversely proportional to E_0, it follows that at energies of the order of 100 MeV these two processes begin to compete and nuclear reactions must be taken into account. They result in the decay of a nucleus into several fragments causing further ionization of the target and the displacement of its atoms.

The interaction of neutrons with nuclei is due to nuclear forces. Therefore, at all energies the cross sections for the elastic and inelastic interaction of neutrons with nuclei are comparable and if the interaction resonances occur, the inelastic scattering may be much more effective than the elastic process. Nuclear reactions are particularly important in the case of irradiation with thermal and slow neutrons, which are generally incapable of displacing atoms in a medium. The cross sections and products of nuclear reactions differ greatly for different atoms and isotopes. Examples of some of the reactions are listed in Table 3.3.

The elastic scattering of photons by nuclei is analogous to the scattering of photons by electrons, known as the Compton process (see Sec. 2.1). However, the interaction cross section is then governed not by the classical electron radius ($r_0 = e^2/mc^2$), but by an analogous nuclear quantity

$$r_0' = \frac{Z^2 e^2}{Mc^2} = r_0 Z^2 \frac{m}{M} \approx 10^{-3} r_0 Z.$$

Since $r_0' \ll r_0$, this process can be ignored compared with the Compton scattering.

The transfer of momentum and energy from a photon to a nucleus occurs also if an electron–positron pair occurs in the field of the nucleus (see Sec. 2.1). This process may be principal for the displacement of nuclei in the case of irradiation of thin samples with high-energy γ rays when the mean free path of electrons and positrons is greater than the thickness of the target.[17]

Resonant absorption and emission of photons by atomic nuclei known as the Mössbauer effect is frequently used in modern physics. In this case the interaction cross section is of the order of λ^2 (where λ is the photon wavelength), which in the case of x rays amounts to 10^{-16}–10^{-17} cm^2. The absorption or emission of a photon results in the acquisition of a recoil momentum $p = \hbar\omega/c$ by a nucleus, which corresponds to a recoil energy $T = p^2/2M = \hbar\omega/2Mc^2$. Therefore, the emission and absorption lines are separated by an energy $2T$ (a photon of energy $\hbar\omega + T$ is absorbed and a photon of energy $\hbar\omega - T$ is emitted). If the absorbing and emitting atom are in a solid and $\hbar\omega$ lies in the x-ray part of the spectrum, the recoil energy is negligible because M is the mass of the whole solid (the minimum energy which a single atom can acquire is equal to the phonon energy, i.e., it is of the order of 10^{-2} eV), which results in the Mössbauer effect.

3.2. Secondary processes

The processes discussed in Sec. 3.1 create primary knocked-out atoms energy T in an irradiated substance. If $T \geqslant E_d$, an atom leaves its equilibrium position. Its subsequent behavior depends on the value of T. At high T the knocked-out atom may displace other target atoms from their positions. The minimum energy required for this purpose can be estimated using Eq. (3.2), which in the case of a simple substance $(M = M_1)$ gives $T_{min} = 2E_d$.

We shall introduce the concept of a cascade function $v(T)$ representing the average number of the displaced atoms formed as a result of one primary displacement. It follows from the above discussion that if $T < 2E_d$, then this function is $v = 1$. We can estimate $v(T)$ for higher values of T in the same way as the number of electron–hole pairs created as a result of ionization of the electron subsystem (Sec. 2.5), by making the assumption

$$v(T) = T/E', \tag{3.13}$$

where E' denotes the average energy lost in the displacement of one atom. Calculations carried out using a simple model (Kinchin–Pease model of elastic pair collisions of hard spheres without any spatial correlation of their positions) give the result $E' = 2E_d$, so that instead of Eq. (3.13) we now have

$$v(T) = T/2E_d. \tag{3.14}$$

According to other models the energy E' can have other values, but they do not differ too much (for example, a computer modeling reported in Ref. 122 gives $E' = 2.13E_d$).

If G_d is the rate of induced primary displacements, then the total rate of the induced displacements can be estimated from

$$G_c = G_d v(T) = \frac{G_d T_d}{2E_d} \approx \frac{1}{2} G_d \ln \frac{T_m}{E_d}. \tag{3.15}$$

Figure 3.2 is an example of a defect-formation cascade induced in a solid by a fast incident particle.

This description of the displacements of atoms in irradiated solids is greatly simplified and is of limited validity. This is due to the fact that at high incident-particle energies the model of pair collisions does not work: a knocked-out atom begins to interact with a neighboring atom before the primary particle leaves the collision region and the free range of the knocked-out atom undergoing secondary collisions approaches the interatomic distance. Moreover, not only new defects are created, but also the existing ones recombine (see Sec. 3.4). These circumstances have led to attempts to provide phenomenological models for estimating the consequences of the primary and secondary displacements considered together.

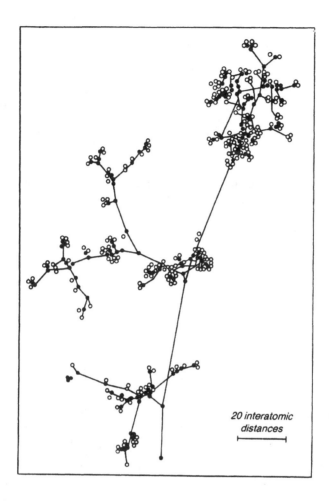

FIG. 3.2. Vacancies (open circles) and interstitial atoms (black dots) created in Ge by a primary displaced atom of 10^4 eV energy.[136]

A basically simple approach is to consider the thermodynamics of the affected part of a crystal allowing for its volume, specific heat, thermal conductivity, and the amount of energy transferred to this region. This model is known as the thermal spike model. It is assumed that the energy of a primary particle Q transferred to a substance is liberated instantaneously in the form of the thermal energy within a small region of a continuous medium and then spreads out in accordance with classical laws. Estimates indicate that the characteristic energy exchange time for the nuclear subsystem ($\approx 10^{-12}$ s) is shorter than the time for the exchange of energy between the nuclear and electron subsystems ($\approx 10^{-11}$ s). Consequently, we can ignore the electron thermal conductivity (even in the case of metals); it is

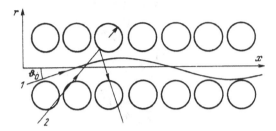

FIG. 3.3. Paths of channeled (1) and unchanneled (2) particles in a crystal (ϑ_0 is the angle of intersection of a path with the channel axis).

sufficient to allow only for the lattice thermal diffusivity D^*, which in the case of many solids is characterized by a value of 10^{-3} cm^2/s.

Under these assumptions the process of radiation damage can be described by the heat conduction equation

$$\frac{\partial^2 T}{\partial r^2} = \frac{1}{D^*}\frac{\partial T}{\partial t},$$

which in the spherical symmetry case has the solution

$$T(r,t) = T_0 + \frac{Q}{(4\pi D^* t)^{3/2} h\rho} \exp\left(-\frac{r^2}{4D^* t}\right), \qquad (3.16)$$

where T_0 is the temperature before irradiation; h is the specific heat; ρ is the density of matter.

Fast melting and subsequent solidification occur in that part of a condensed substance where T exceeds the melting point. It is natural to expect that a part of a crystal where such melting and solidification occur is characterized by a somewhat different distribution of atoms than that present initially. Some additional processes can occur also in the vicinity of a thermal spike where the temperature does not reach the melting point. For example, plastic deformation and acceleration of diffusion can take place there.

Estimates based on Eq. (3.16) show that if the source of a thermal spike in copper is a knocked-out atom of energy 300 eV, then in a time of 5×10^{-12} s (representing about 30 lattice vibration periods) a region 3 nm in diameter melts and this region contains 1100 atoms; after 2×10^{-11} s in a region of radius 6 nm the average temperature amounts to 150 °C (Ref. 51).

At ultrahigh energies of the incident particles in solids we can expect effects fundamentally different from those discussed above. If the energy transferred by the first collision is so high that it results in the approach of several nuclei to distances of the order of the radius of action of nuclear forces, the result may be a fireball consisting of nuclear matter. Such fireballs explode creating shock waves consisting of nucleons and pions. These phenomena have been observed in NaF crystals bombarded with accelerated neon nuclei.[126]

The experimental results of investigations of tracks and ranges in crystals are generally in good agreement with theoretical models allowing for the ionization losses and cascade enhancement in the number of the displaced atoms. However, detailed investigations have shown that some of the displaced atoms have ranges much higher than those expected. It is found that the paths of such atoms pass along open channels which appear in any periodic crystal lattice and are located between close-packed rows of planes of atoms. This effect is known as the particle channeling.

The nature of motion of an atom in a channel can be determined by considering the simple example of a two-dimensional channel formed by two atomic planes (Fig. 3.3). The potential of a particle in a channel $U(r)$ can be calculated as the sum of the Born–Mayer potentials and of the screened Coulomb potential of several atoms closest to the particle. It is found that at low values of r the result can be approximated satisfactorily by a quadratic function of r, i.e., by the harmonic potential $U(r) = \alpha r^2$ (here, α is a constant of a given substance). In a field created by this potential the particles experience a force $F(r) = -2\alpha r$ and the motion of a particle along the coordinate r is described by the equation

$$\frac{d^2 r}{dt^2} + \frac{2\alpha r}{M_1} = 0,$$

where M_1 is the mass of a particle. The solution of this equation is a harmonic vibration of a particle with a period

$$\tau_0 = \frac{2\pi}{\omega_0} = 2\pi \left(\frac{M_1}{2\alpha} \right)^{1/2}. \tag{3.17}$$

The velocity of a particle along a channel is $v = (2E/M_1)^{1/2}$, so that one transverse vibration is completed in the time in which an atom travels a distance $\lambda = v\tau_0 = 2\pi (E/\alpha)^{1/2}$, and the equation of motion along the coordinate r is of the form

$$r = r_c \sin \left[\left(\frac{\alpha}{E} \right)^{1/2} x \right], \quad r_c = \vartheta_0 \left(\frac{E}{\alpha} \right)^{1/2}, \tag{3.18}$$

where ϑ_0 is the angle at which the particle path intersects the channel axis and r_c is the effective width of the channel.

It follows from Eq. (3.18) that a channeled particle exhibits sinusoidal vibrations about the channel axis. An increase in E and ϑ_0 increases the amplitude r_c of these vibrations. When r_c is large, a channeled particle begins to interact with the individual atoms in the channel wall and the channel collapses because one of the atoms in the channel is displaced from its normal position (Fig. 3.3). Therefore, there are certain maximum values $\vartheta_0(E)$ and $E(\vartheta_0)$ which limit channeling. They are related by $\vartheta_{0c} = (U_0/E)^{1/2}$, where U_0 is the energy depth of the channel. Typical values of ϑ_{0c} and U_0 are about $2°$ and 30 eV, respectively.

The channeling energy has also a lower limit. This is due to the fact that a reduction in E reduces also λ. If in the course of deceleration of a particle the value of λ approaches the interatomic distance in the wall of a channel, a resonance is established between the walls and vibrations of the particle, and the particle path terminates by a series of strong collisions with the walls of the channel and destruction of the latter. When atoms of copper travel in solid copper this occurs at $E \approx 300$ eV.

This simple classical approximation is invalid in the case of the channeling of light charged particles such as electrons and positrons. In their case the channeling

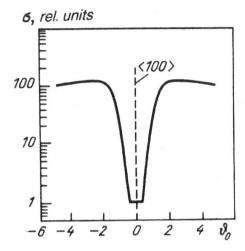

FIG. 3.4. Efficiency of the backscattering of 1-MeV protons by the surface of a tungsten crystal plotted as a function of the angle of incidence (in degrees) relative to the $\langle 100 \rangle$ axis.[67]

and diffraction angles are comparable, and a detailed investigation of the channeling characteristics should be made allowing for the quantum effects.

The channeling influences the formation of defects in crystals in two ways. Firstly, a reduction in the probability of collisions of particles with atomic nuclei and the loss of the bulk of the energy of a particle in small portions reduces the efficiency of defect formation. Secondly, as pointed out already, the average range of the primary and secondary particles increases.

The most popular method for experimental investigation of the channeling is determination of the efficiency of backscattering of fast incident particles by crystalline surfaces. At angles of incidence satisfying the channeling conditions the efficiency of this process falls (Fig. 3.4).

The channeling effect has had a large number of interesting practical applications. We shall mention only some of them.

Interstitial atoms block channels in a crystal and, consequently, they reduce the channeling probability. This makes it possible to investigate the lattice defects by a study of the efficiency of channeling of fast particles.

The angular distribution of channeled particles emitted by crystals contains information on the crystal lattice structure. A structural analysis method based on this circumstance utilizes particles created in a crystal as a result of nuclear reactions and is known as the shadow method.

Since positively charged channeled particles are located far from the nuclei of the crystal atoms, channeling reduces the probability of nuclear reactions involving such particles. In the case of negatively charged particles there should be an increase in the probability.

Bent single crystals can be used to control the paths of channeled particles. For example, bending of a silicon single crystal 20 mm long through an angle of 26 mrad (1.5°) made it possible to deflect some of 8-GeV protons crossing a crystal through the same angle, compared with the initial direction of flight. A magnetic field which can achieve this effect would have to be of 72 T intensity, tens of times

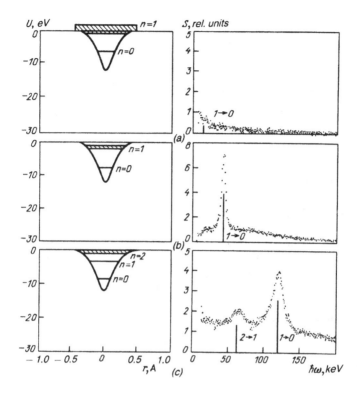

FIG. 3.5. Effective potential with energy levels and emission spectrum (with an indication of the dominant transitions) obtained by excitation with electrons of energies 16.9 MeV (a), 30.5 MeV (b), and 54.4 MeV (c) channeled in diamond crystals after entering through the (100) surface.[101]

higher than the fields used in accelerators. It is interesting to note also that in the case of bent crystals the value of ϑ_{0c} is greater than for those which are not bent.[4]

Channeled particles considered from the point of view of transverse motion relative to the channel axis are in a potential well and their energy can have only discrete values separated by intervals of several electron volts (Fig. 3.5). Transitions between these energy levels result in the emission of photons from particles. Since an emitting particle is traveling at a high velocity along the channel axis, the radiation is directed along this axis and the photon energy is shifted, in accordance with the Doppler effect, toward x-ray and γ-ray parts of the spectrum. The emission spectrum consists of quasicharacteristic lines and the energy and number of these lines increase on increase in E (Fig. 3.5).

An interesting effect is the excitation of the electron subsystem of channeled particles in the case when the frequency v/a of the interaction of these particles with the crystal atoms (a is the lattice constant) is in resonance with the frequency of any specific electron transition. This results in transitions characterized by energies $\Delta E = k\hbar v/a$ ($k = 1,2,3,...$). An example of such transitions is the experimentally observed ionization of the N^{6+} ions (i.e., the transition $N^{6+} \rightarrow N^{7+}$) channeled

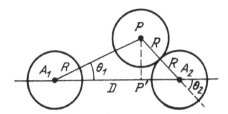

FIG. 3.6. Collision of two identical particles.

along the $\langle 111 \rangle$ axis in gold, which exhibits resonances at $E_0 = 33.07$, 21.17, and 14.70 MeV, corresponding to $k = 4$, 5, and 6 (Ref. 75).

Low-energy atomic collisions in ordered solids give rise to an additional typical crystal effect which is the focusing of collisions along close-packed rows of atoms.

We shall consider a collision of two elastic hard spheres of radius R with their centers initially located at points A_1 and A_2 separated by a distance D (Fig. 3.6). We shall assume that one of them has a momentum along the direction A_1P, which makes an angle θ_1 with the line A_1A_2. At the moment of collision its center is at the point labeled P. A collision imparts a momentum to the second atom along the PA_2 direction and this direction makes an angle θ_2 relative to the A_1A_2 line. At low values of θ_1 and θ_2, we have

$$A_1P \approx D - 2R, \quad PP' \approx (D - 2R)\theta_1 \approx 2R\theta_2,$$

which can be used to calculate what is known as the focusing parameter

$$f = \frac{\theta_2}{\theta_1} \approx \frac{D - 2R}{2R} = \frac{D}{2R} - 1. \tag{3.19}$$

After n collisions, the collision angle becomes

$$\theta_n = f^n \theta_1.$$

Hence, it is clear that when the condition $D < 4R$ is satisfied, we have $f < 1$ and $\theta_n < \theta_1$, i.e., the direction of motion of each next atom in a chain of collisions approaches the general direction of the row of atoms, i.e., the collisions become focused (Fig. 3.7). In the case of such focusing both the momentum and energy, transferred to any atom in a crystal, may propagate over long distances from the location of the primary collision. If the energy transferred along a chain of atoms exceeds the energy needed to replace one atom with another, then a chain of substitution collisions propagates along a crystal and this is known as a dynamic crowdion, which is a part of a chain with a higher density (Fig. 3.7). If the kinetic

(a)

(b)

FIG. 3.7. Chain of focusing collisions (a) and a dynamic crowdion (b).

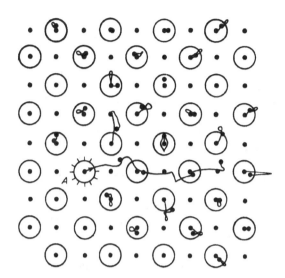

FIG. 3.8. Paths of atoms in a crystal after a kinetic energy of 40 eV is imparted to an atom *A* along the direction identified by the arrow.[87]

energy transferred by scattering to the neighboring atomic rows is less than the substitution energy, a crowdion stops and forms an interstitial atom far from a vacant site created by the primary collision. If a chain of substituting collisions reaches the surface of a crystal, then an atom escapes from a crystal along this chain.

Figure 3.8 gives the results of computer modeling of the motion of atoms in a crystal after one of its atoms receives a specific energy and a specific momentum. This figure demonstrates clearly the occurrence of focusing collisions and displacements along close-packed rows of atoms.

It follows from the above discussion that focusing collisions play the same role in the formation of defects in a crystal as the channeling effects. In a given crystal such collisions occur mainly along the direction with a minimum value of D.

In the model of pair collisions of identical hard spheres the quantity R depends on the transferred kinetic energy T and the interatomic potential $U(r)$. If we take $2R$ to be the minimum distance of approach between two colliding atoms, then in the case of head-on collisions the approximation of the interatomic potential by the repulsive Born–Mayer potential gives

$$\frac{T}{2} = U(2R) = A \exp\left(-\frac{2R}{b}\right), \quad R = \frac{b}{2} \ln \frac{2A}{T},$$

where A and b are constants of values that depend on the nature of the colliding pair of atoms and in many cases amount to 10^3–10^5 eV and 0.03 nm, respectively. Focusing collisions occur if

$$D < 4R = 2b \ln \frac{2A}{T},$$

which means that focusing is possible if T is less than a certain critical value known as the focusing energy

$$E_F = 2A \, \exp \left(-\frac{D}{2b} \right). \tag{3.20}$$

Typical values of E_F are 10–50 eV.

A major role in the propagation of focusing collisions is played by the interaction between moving atoms and the neighboring atomic rows. This interaction dissipates some of the energy of a crowdion. However, the neighbors can also make for the collision process: an atom escaping as a result of defocusing of collisions (if $f > 1$) from its row returns to that row because of the interaction with the atoms in the neighboring rows. This effect is known as assisted focusing and may result in focusing of collisions also when $E > E_F$.

Collisions of atoms with different masses result in partial transfer of the kinetic energy [see Eq. (3.2)]. The effect is rapid dissipation of the energy of crowdions moving along corresponding directions in crystals of compounds.

The channeling and collisional focusing effects appear most strongly when the periodicity of the crystal lattice is perfect. Thermal lattice vibrations which disturb these conditions cause additional dissipation of the energy of channeling particles and of crowdions. An increase in temperature shortens the channels and crowdions.

3.3. Perturbation of the nuclear subsystem by electronic transitions

In the previous sections we assumed that the electron and nuclear subsystem of a substance interact independently with ionizing radiations. This assumption is a direct consequence of the adiabatic approximation (see Sec. 1.2). In this approximation the change in the electron subsystem energy ΔE_e behaves as a change in the potential energy of the nuclear subsystem. If the excitation of the electron subsystem is of local (non-Bloch) nature, it should give rise to a force $F = d(\Delta E_e)/dR$, which tends to alter the spatial distribution of the nuclei. Consequently, the nuclear subsystem acquires a kinetic energy which can be estimated in the harmonic approximation from

$$\Delta E_t = \frac{1}{2M} \left| \int_{-\infty}^{\infty} F(t) \exp(-i\omega_0 t) dt \right|^2, \tag{3.21}$$

where M is the mass of the nuclei in the region interacting with electron radiation and ω_0 is the effective vibration frequency of the nuclei.

We shall consider the situation when an electron excitation of a polyatomic system is localized for a time τ_e in the region of one atom, i.e., this interaction has a radius of the order of the interatomic distance a. We shall assume that if the kinetic energy ΔE_t accumulated by an atom exceeds a certain threshold value E_d, the atom is displaced from its equilibrium position to a distance a, which means that a structure defect appears. If we assume that $F(t) = (\Delta E_e/a) \exp(-t/\tau_e)$ and $E_d = \frac{1}{2} a^2 \omega_0^2 M$, we find from Eq. (1.21) that the condition for the formation of a defect is

$$\Delta E_e \geqslant E_d \frac{(1 + \omega_0^2 \tau_e^2)^{1/2}}{\omega_0 \tau_e}, \tag{3.22}$$

which in the limiting cases of long- and short-lived electron excitations reduces to

$$\Delta E_e \gg E_d \quad \text{if } \tau_e \gg \omega_0^{-1}, \tag{3.23}$$

$$\Delta E_e \gg \frac{E_d}{\omega_0 \tau_e} \quad \text{if } \tau_e \ll \omega_0^{-1}. \tag{3.24}$$

If the condition (3.23) is satisfied, a potential displacement takes place, whereas in the case of the condition (3.24) we can expect an impact displacement of the investigated atom.

Since the harmonic nature of the motion of an atom is lost when it moves away a distance a from its equilibrium position, the above expressions provide only qualitative estimates. They reflect the trend that if an electron excitation has a sufficient energy, the probability of formation of a lattice defect at the expense of this energy rises on increase in the time of localization of an excitation in the vicinity of one atom.

For the majority of condensed systems the frequency ω_0 is of the order of 10^{13} s^{-1}. When the condition (3.23) is satisfied, we must ensure that $\tau_e \gg 10^{-13}$ s. Using the Nernst–Einstein relationship $\mu_e/D_e = e/k_B T$ and the definition $D_e = a^2/2\tau_e$ [see Eq. (1.35)], we can demonstrate that this condition corresponds to an excitation mobility μ_e below 1 cm^2 V^{-1} s^{-1}. Such low mobilities of electron excitations are encountered in various molecular systems, in disordered solids, and also in crystals characterized by a strong electron–phonon interaction.

We must also refine the meaning of the threshold energy of the formation of a defect E_d. It is the energy necessary to displace an atom from a site to an interstice, and it is governed by the energy of the atom in the most close-packed part of its path. This energy depends on the degree of relaxation of the environment of the atom, i.e., on the time a/v during which an atom is located in this region (v is the velocity of the atom being removed) decreasing on reduction in the kinetic energy of the atom in question. If $a/v \gg \omega_0^{-1}$, i.e., if an atom moves away slowly, the lattice around the newly formed defect is at each moment almost in an equilibrium state and then E_d is close to the equilibrium energy of the formation of a defect which is 1–4 eV for many solids. However, if $a/v \ll \omega_0^{-1}$, the lattice around a newly formed defect does not relax in the available time and E_d assumes high values close to those encountered in the formation of defects by the direct interaction of radiation with the nuclear subsystem of matter (Sec. 3.1).

In accordance with the generally accepted classification of the forces of interaction of atoms in molecules and in condensed media, we can divide the elementary mechanisms of perturbation of the nuclear subsystem (as a result of electron transitions) into two classes: Coulomb and Pauli.

The primary and secondary ionization processes in an irradiated substance may create doubly or multiply ionized atoms (see Sec. 2.4). An electrostatic force appears in the vicinity of these atoms and it is capable of destabilizing the region around the ionized atom. This force can be estimated from

$$F(t) = \frac{Z_1 Z_2 \alpha e^2}{\varepsilon(r) r^2(t)}, \tag{3.25}$$

where $Z_1 e$ and $Z_2 e$ are the interacting charges; α allows for the environment-

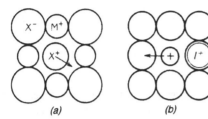

FIG. 3.9. Schematic representation of the displacement to an interstice of a positively charged anion in an ionic crystal (a) and of a positively charged atom located near a positively charged impurity I^+ in a semiconductor (b).

induced polarization and Madelung corrections to the pure Coulomb potential; $\varepsilon(r)$ is the effective permittivity at a time-dependent distance r between the interacting charges.

There are many examples of such a radiation-induced Coulomb instability (RICI) of molecules. For example, it is reported in Ref. 95 that the Coulomb repulsion between charged fragments of benzene molecules ionized by electrons causes them to fly apart, whereas the results given in Ref. 71 show that a Coulomb explosion occurs in CH_3I molecules after the photoionization of the L shell of iodine: $CH_3I \rightarrow C^{2+} + I^{5+} + 3H^+ + 10e$. Other examples are decay of doubly charged microclusters $(Pb)_n^{++}$, $(NaI)_n^{++}$, and $(Xe)_n^{++}$ into two fragments if n is sufficiently small ($n_{max} = 20$–52, depending on the chemical binding force).[123]

The possibility of the RICI in solids was first pointed out in Ref. 133, where it was postulated that if an anion in an ionic crystal loses simultaneously two or more electrons and thus acquires a positive charge, then its position at a regular lattice site becomes unstable because of the repulsion by the nearest cations and it may be displaced to an interstitial position with the aid of the lattice vibrations (Fig. 3.9). Subsequent detailed theoretical and experimental investigations have shown however that this Varley defect formation mechanism is very ineffective. The main reason for this is the limited lifetime of a multiply charged anion (it is of the order of 10^{-12}–10^{-13} s), so that the condition of Eq. (3.22) is not satisfied: in a close-packed highly symmetric lattice consisting of ions of variable sign an anion cannot acquire the energy and momentum necessary for displacement to an interstice during its lifetime.[80]

However, the RICI may result in defect generation when the environment of an ionized atom is asymmetric, so that right from the beginning there is a preferential direction of its motion. Two types of such a situation are known.

One of them occurs on the surface of an ionic crystal where, because of the low coordination number and the possibility of electron emission, all the channels for the neutralization of an ionized anion are less effective and the anion can escape readily from a crystal along the normal to its surface. The feasibility of this process had been demonstrated experimentally for NaF crystals from which Na^+ and F^+ ions are emitted after the photoionization of the K shell of Na^+ (Ref. 115).

Another example of such processes is the impurity-ionization mechanism of the formation of defects in doped semiconductors,[30,102] which reduces to expulsion of an ionized atom (with a charge Z_1e) from a regular site as a result of repulsion from a nearby charged impurity (Fig. 3.9). The formation of a defect in accordance with this mechanism is possible inside a sphere of radius $R_a \approx Z_1Z_2e^2/\varepsilon_0E_d$ around an impurity atom with a charge Z_2e; this sphere contains N_a of atoms of the host

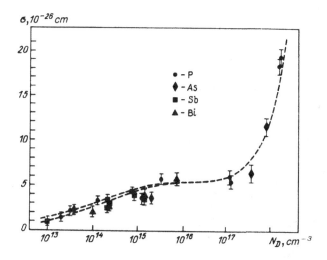

FIG. 3.10. Dependence of the cross section of formation of defects in Ge irradiated with ^{60}Co γ rays on the group V donor concentration N_D (Ref. 23).

substance. Therefore, the cross section for the formation of a defect is proportional to $N_a\Sigma_j\sigma_j$, where σ_j is the ionization cross section of the jth electron shell of an individual atom. Experiments have demonstrated that this defect formation mechanism plays an important role in crystals of Ge (Fig. 3.10), InSb, and diamond containing charged impurities in sufficiently high concentrations (in excess of 3×10^{17} cm^{-3} in the case of Ge).

A collective variant of the RICI applies to insulators and semiconductors, when a high-energy particle characterized by large linear energy losses ionizes crystal within a thin cyclinder from which electrons are removed temporarily and which therefore contains a large number of positive ions. Many of these ions may leave their regular positions before neutralization and they may form a track consisting of defects. This track formation mechanism may also be active on the surfaces of thin metal films depleted of electrons as a result of their emission.[85]

In the Pauli mechanisms of perturbation of the nuclear subsystem in matter the driving force appears between overlapping electron shells of atoms or ions as a result of the Pauli principle. In the case of molecules this force may be due to the removal of an electron from a bonding orbital or its transition to an antibonding orbital or even as a result of the capture of an additional electron by an antibonding orbital (this is known as the dissociative capture).

A very frequent reason for the decay of excited molecules is their predissociation. In quantum mechanics this dissociation is due to a nonradiative transition of a system from a discrete (bound) state to free states in the continuous spectrum under the action of some perturbation ΔU, which is usually described by the non-adiabatic potential representing the influence of the kinetic energy of the nuclei on the electron wave functions $\Psi(r;R)$ (see Sec. 1.2). The matrix element $M_{f0} = \int\varphi_f^*(R)\Delta U(R)\varphi_0(R)dR$ of such a transition is usually calculated on the as-

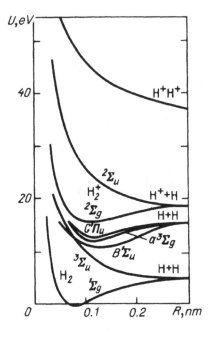

FIG. 3.11. Dependences of the potential energy on the interatomic distance obtained for different electron states of the H_2 molecule.[18]

sumption that ΔU is constant; the functions φ_f and φ_0 are then the wave functions of one-dimensional motion in a homogeneous force field which can be analyzed semiclassically. The result is the Landau–Zener formula, according to which the transition probability P_d is

$$P_d = \exp\left(-\frac{2\pi(\Delta U)^2}{\hbar v_c |F_2 - F_1|}\right), \tag{3.26}$$

where $v_c = (dR/dt)|_{R_c}$ is the velocity at which this system crosses the point of intersection of the potential curves of the investigated states R_c; F_1 and F_2 are the slopes of the straight line approximating the potential curves at this point described by $F_i = [dU_i(R)/dR]|_{R_c}$. It follows from Eq. (3.26) that, in particular, the probability of a nonradiative transition increases on reduction of the difference between the slopes of the potential curves of the initial and final states in the region of the transition.

A more detailed analysis must allow also for the vibrational structure of the states participating in the transition. This can be done, for example, by expressing the wave functions in terms of the Airy functions in which the dependence on the vibrational number is introduced formally.[48]

The most frequent type of perturbation of the structure of a molecule by this mechanism is fragmentation. Since each excited electronic state of a molecule has its own characteristic decay channels, the overall fragmentation pattern can be very complex even in the case of relatively simple molecules. For example, the H_2 molecule has the simplest electronic structure but even then it is found that the $^3\Sigma_u^+$ state gives rise to two neutral H atoms, the $^2\Sigma_u$ state of the H_2^+ ion yields an atom of H and an H^+ ion (proton), whereas the singlet states $^1\Sigma_u^+$ and $^1\Pi_u$ are very

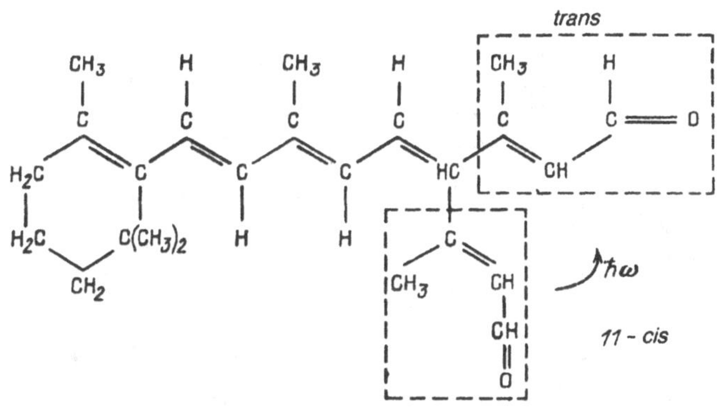

FIG. 3.12. Structural formula of the retinal molecule.

likely to decay radiatively (Fig. 3.11). The fragmentation products of excited water molecules include H_2O^+, OH^+, H^+, O^+, H, OH, and H_2.

The ionization and excitation of organic molecules is also followed by a great variety of decay processes. Molecules of this kind have a wide range of bonds and those most likely to break are the C—H, C—C, and C—O. In the case of hydrocarbons the most probable primary reaction is of the $C_nH_{2n+2} \rightarrow C_nH_{2n} + H_2$ type [for example, in the case of propane this may be the $(C_3H_8)^* \rightarrow C_3H_6 + H_2$ reaction].

One of the possible reactions of a complex molecule to excitation or ionization is the change in its structure by reorganization of the system of bonds, i.e., isomerization of molecules. In the case of degenerate excited states the isomerization involves lifting of the degeneracy (Jahn–Teller effect). For example, the primary process ensuring human vision is the *cis→trans* photoisomerization of the retinal molecule in the retina of the eye (Fig. 3.12).

Such a transition occurs in ≈ 10 ps and its quantum efficiency is 0.5–0.7 when the excitation is provided by photons from the visible part of the spectrum, but the quantum efficiency is much lower (a few percent) when high-energy charged particles provide the excitation. The energy of *cis*-retinal in its relaxed excited singlet state is 0.5 eV higher than the energy of *trans*-retinal, which stimulates the isomerization. In the ground state of the molecule the situation is reversed, so that the *cis* form is restored after decay of the excitation.[132]

The efficiency of these processes depends, in principle, also on the vibrational state of a molecule. The most convenient method for a deliberate change in the mode composition of the vibrational excitation of molecules is exposure to infrared laser radiation of frequency equal to the frequency of some vibrational mode. There are some interesting examples of a considerable increase in the reaction rate if one of the reagents is transferred from the zeroth vibrational level to the first one ($K + HCl \rightarrow KCl + H$; $Sr + HF \rightarrow SrF + H$, etc.). It has been shown[1] that even the nature of fragmentation of molecules by electron impact can depend on the vibrational state.

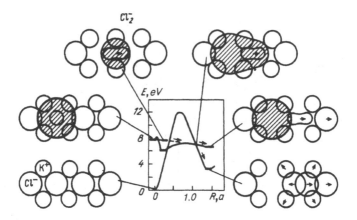

FIG. 3.13. Dissociation of an exciton into lattice defects in a KCl crystal. The shaded region represents approximately the zone of localization of a weakly bound electron. The center of the figure shows the energy scheme of the process plotted on the basis of Ref. 58.

The Pauli mechanisms of modification of the nuclear subsystem in solids exhibiting a strong electron–phonon interaction may be encountered when the condition (3.22) is satisfied. Such a situation occurs if the energy liberated by the recombination of electrons and holes or by the annihilation of excitons is not less than the energy required for the formation of a lattice defect, i.e., if the condition $E_g \geqslant E_d$ is obeyed.

The most reliably established and thoroughly investigated case of formation of a lattice defect as a result of nonradiative decay of excitons is the process which occurs in alkali halide crystals.[30,38] Direct experiments have shown that the decay of an exciton and the recombination of an electron with self-localized hole in KCl, KBr, and similar crystals creates neutral and charged pairs of anion Frenkel defects (F-center–H-center and α-center–I-center pairs—see Sec. 3.4) in a time of 10^{-11} s.

We shall consider this process in the specific case of KCl crystal (Fig. 3.13). The equilibrium energy of a pair of F-center–H-center defects is about 7 eV and the internal energy of a free exciton is about 8 eV. Such an exciton has a tendency to become self-localized by analogy to the self-localization of a hole (see Sec. 2.5): its hole component forms a local covalent bond between the two nearest anions and this creates a quasimolecule X_2^- spread over two lattice sites (Fig. 2.31). A self-localized exciton may be in various electronic states and the energy of these states is 6–8 eV relative to the energy of the ideal lattice. Therefore, the decay of an exciton into an F–H pair is possible for energy reasons if the exciton is in an unrelaxed or partly relaxed state. The decay itself is due to a transition of the hole component of an X_2^- exciton from two sites to one, which is equivalent to the formation of an interstitial halogen atom in a relaxed state, known as the H center. The latter may move away from the point of its generation by interstitial diffusion or in the form of a crowdion. The weakly bound electron in the exciton remains in the field of the resultant anion vacancy and it forms an F center at the point where the exciton decays. If we regard a self-localized exciton as an excited molecule $[(M^+)_2 X_2^{2-}]^*$ (where M^+ is an alkali metal ion), then the process can be regarded as the isomerization of this molecule due to the pseudo-Jahn–Teller effect: relaxation of an excited molecule may give rise to a situation in which odd vibrations of a molecule (i.e., the displacements of X_2^- as a whole relative to the M^+ ions along the [110] direction) mix the odd (excited) and even (ground) states of the molecule, so that the molecule becomes unstable against these vibrations (i.e., X_2^- is displaced from the position between the two M^+ ions). If the efficiency of this process is described by Eq. (3.26), a reasonable agreement with the experimental results is obtained if we assume that $v_c = 1.2 \times 10^5$ cm/s, $|F_2 - F_1| \approx 10^8$ eV/cm, and $\Delta U \lesssim 0.04$ eV (Ref. 99).

When the H center moves from the F center by a distance exceeding a, the energy of such an F–H pair of centers is higher than the energy of an α–I pair of charged centers (Fig. 3.13). In the range of distances where these energies are approximately equal it is highly probable that an electron localized mainly near the anion vacancy is transferred to the H center and this gives rise to a pair of the α–I centers (Fig. 3.13). Therefore, the products of the defect-forming decay of excitons include both F–H and α–I pairs.

In those semiconductors and insulators in which low-energy excitons are insufficiently energetic to create defects in the regular parts of the lattice, new defects may be created as a result of the combined influence of the thermal fluctuations and electron excitations or of the motion and transformation of the existing defects.[14] The energy barriers for these processes may decrease considerably or vanish completely in the presence of electron excitations. This results in radiation-stimulated or radiation-enhanced diffusion, which is the cause of a considerable modification of the defect structure of semiconductors and of degradation (aging) of many semiconductor devices.

In the case of dense ionization of large microregions of matter (for example, in tracks of charged particles[13] or in the field of laser radiation[40,131]) we can expect very complex effects which can disturb the structure or even fracture crystals. This may happen for a great variety of reasons which are very difficult to identify. This applies to various phase transitions, Coulomb explosions in a certain region, electrical breakdown, thermal shock, and other effects which are difficult to interpret quantitatively.

The transfer of the energy of electron excitations to the nuclear subsystem is least likely in metals: the absence of a band gap and the presence of a large number of free electrons means that any radiation-induced nonequilibrium distribution of charges is rapidly neutralized and the lifetime of electron excitations decreases considerably because of the high efficiency of their Auger decay. Nevertheless, there are some cases when we can speak of the influence of electron excitations on the formation of radiation defects also in metals. It is found that the threshold energy for the formation of defects E_d in W and Al depends on the magnitude of the ionization energy losses $|dE/dx|_{\text{ion}}$ of the incident particles: the higher the value of $|dE/dx|_{\text{ion}}$, the smaller the energy E_d (Ref. 54). The effect is presumably due to the circumstance that in a region of a crystal where the electron subsystem is disturbed from its equilibrium state the energy E_d is less than the usual value.

Selective perturbation of certain molecules in a given mixture of molecules or of

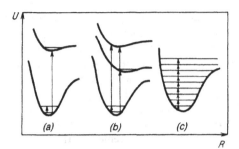

FIG. 3.14. Selective photoexcitation of molecules: (a) one-stage excitation of electronic or vibrational states; (b) two-stage excitation of electronic states via vibrational or electronic states; (c) multiphoton excitation of vibrational states.

particular sites in the crystal lattice in solids of complex composition is of considerable interest. This can be achieved, in principle, in those cases when nonidentical electron excitations result preferentially in specific perturbations of the nuclear subsystem.

A Coulomb instability of a molecule or a solid can be induced selectively by utilizing the difference between the ionization energies of the inner electron shells of different atoms or of the same atoms but in different chemical environments, and also by making use of the fact that after the ionization of a specific atom an Auger cascade is confined in practice to this atom (see Sec. 2.4). For example, it has been shown that fragmentation of the acetone molecule after the photoionization of the K shell of carbon occurs preferentially in the region of the C atom which has been ionized and that the fragmentation products depend on the primary excited state of the molecule.[77]

Powerful techniques for selective perturbation of molecules are provided by what is now known as laser chemistry. These techniques utilize the high intensity and monochromaticity of laser radiation, and are similar to the selective photoionization of atoms discussed in Sec. 2.1 (Ref. 37).

The main process specific to laser chemistry is multistage photoionization of molecules (Fig. 3.14). The high monochromaticity of laser radiation, selection of the energy of the incident photons in accordance with the spectrum of the electronic–vibrational states of the molecules of interest to us, and the utilization of the differences between the energies of the states, their lifetimes, and degrees of excitation (ionization) can ensure selectivity of the excitation or ionization of bonds or molecules.

Selective laser chemistry has important applications in isotope separation.[28,37] For example, in the case of tetracene ($C_2N_4H_2$) the concentrations of the molecules containing the ^{13}C and ^{15}N isotopes can be increased by a factor of about 10^4 compared with the original mixture. In the case of CF_3I and CF_3Br it is possible to create conditions under which separation of the molecules with the ^{13}C isotope occurs at a rate of 1 g/min. Selective laser ionization has been used successfully to prepare pure AsCl and SiH_4.

One of the selective methods of laser chemistry is the resonant multiphoton excitation of specific vibrational modes of molecules (Fig. 3.14). The mode selectivity can be achieved only if the duration of multistage excitation of a given mode is less than the time needed for intramolecular redistribution of the vibrational energy between all the modes of a given molecule (about 10^{-12} s). The success is

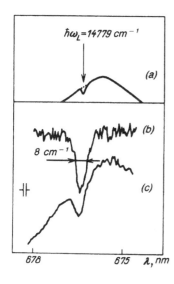

$\hbar\omega_L = 14779$ cm^{-1}

(a)

(b)

8 cm^{-1}

(c)

678   675 λ, nm

FIG. 3.15. Photochemical burning of a "hole" (dip) in the absorption spectrum of t-Bu$_4$-4-PcH$_2$ in C$_9$H$_{20}$: (a) position of an absorption band with a dip; (b) dip shown on a larger scale; (c) experimentally determined spectrum.[88]

most likely expected in the excitation of high frequency local vibrational modes of molecules. These are particularly the C—H, C—D, and C—T bonds in large organic molecules which, because of the high degree of anharmonicity, are coupled only weakly to other modes. For example, in the case of the naphthalene molecule the energy of the C—H mode is 3000 cm^{-1} and the dissociation energy of the C—H bond is 44 000 cm^{-1}. Therefore, successive absorption of fifteen photons of 3000 cm^{-1} energy should dissociate a naphthalene molecule. Examples of successful realization of this theory are photoisomerization of the H$_2$C=CHCH$_2$NC and CH$_2$=CHCH$_3$ molecules after multiphoton excitation of the C—H bonds[120] and selective multiphoton dissociation of CTCl$_3$ molecules by NH$_3$ laser radiation (12.08 μ).[92]

A special feature of selective laser photochemistry is the burning of a dip ("hole burning") in the spectra of large organic molecules dissolved in solid organic matrices.[49] In view of the great variety of possible positions of impurity molecules in such matrices, their spectra are much wider than the corresponding spectra of free molecules (inhomogeneous widening). If such molecules are exposed to monochromatic laser radiation, it is possible to dissociate molecules at specific positions without affecting those at different positions. Such spatially selective photodissociation creates a narrow (10^{-2}–10^{-3} of the total width of the band) dip or "hole" in the part of the spectrum where transitions occur in dissociated molecules. An example of "hole burning" is shown in Fig. 3.15. There are promising applications of this effect in information storage.

Similarly, laser excitation of near-defect local vibrations can stimulate strongly and selectively the diffusion of defects in semiconductors and insulators.[116]

3.4. Defects in crystals

As a result of the processes discussed in Secs. 3.1–3.3 radiation structure defects arise in the nuclear subsystem of the matter.

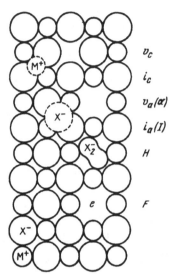

FIG. 3.16. Structure of primary radiation defects in alkali halide crystals.[38]

In crystals the primary defects are vacancy–interstitial atom or ion (v–i) Frenkel pairs. If a solid is a compound, several types of such defect can appear (for example, $v_c + i_c$ and $v_a + i_a$ cation and anion pairs in ionic crystals), as shown in Fig. 3.16.

In principle, such defects may exist in several charge states. For example, anion vacancies can exist in alkali halide crystals without an electron v_a^+ (the upper index indicates the sign of the charge of a defect relative to the regular lattice), frequently called the α or F^+ centers, with one electron ($v_a^+ e$ or v_a^0)—the F centers—or with two electrons ($v_a^+ e_2$ or v_a^-)—the F^- or F' centers; there may be also neutral

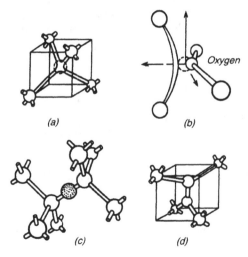

FIG. 3.17. Defect configurations in the diamond lattice: (a) vacancy; (b) complex consisting of a vacancy and an oxygen impurity atom; (c) interstitial atom in the middle of a bond; (d) split interstitial.

(X_i^0 or i_a^0, known as the *H* centers) and charged (X_i^- or i_a^-, known as the *I* centers) states of an interstitial halogen (Fig. 3.16). Neutral and charged interstitial silicon atoms (Si_i^+, Si_i^0, Si_i^-) and vacancies (v^+, v^0, v^- and possibly also v^{2+} and v^{2-}) have been detected in silicon single crystals.

A detailed structure of a defect is determined largely by lattice relaxation around it. For example, in the case of a vacancy in an ionic crystal the nearest neighbors of a defect move away from it (because of the identical signs of the effective charge), whereas the second-nearest neighbors approach it. Such relaxation reduces the energy of a vacancy by about 0.5 eV. In the case of semiconductors we find that reorganization of the covalent bonds occurs near defects and this results in a much more complex equilibrium local configuration of the lattice (Fig. 3.17). A configuration with local covalent bonds may appear also in ionic crystals. For example, an interstitial halogen atom X_i^0 forms a bond with one of the X_a^- anions of the normal lattice characterized by an energy of about 1 eV, so that a molecular halogen ion known as the *H* center (Fig. 3.16) appears at one lattice site $(X_2^-)_a$.

One of the consequences of lattice relaxation around a defect is a change in the volume of a crystal (by an amount Δv) and a local change in the lattice constant (by an amount Δa). The relative changes in the volume $\Delta v/v$ (where v is the volume of a normal site) have values ranging from -0.5 to $+0.5$ for vacancies and 2–3 for interstitial atoms.

In view of the translational symmetry of the lattice, a defect can occupy a large number of equivalent positions separated by energy barriers. Thermal fluctuations help to overcome these barriers so that migration from one site or interstice to another is possible by analogy to hopping of localized electron excitations (Sec. 2.5). The rate of such migration can be represented by the jump frequency

$$P_m = P_0 \exp\left(-\frac{E_m}{k_B T}\right)$$

or the diffusion coefficient

$$D = \frac{a^2}{2n} P_m = D_0 \exp\left(-\frac{E_m}{k_B T}\right), \qquad (3.27)$$

where P_0 is a constant of the order of the effective frequency of the lattice vibrations; E_m is the activation energy of migration, which includes the height of the barrier between equivalent positions of a defect and a lattice as well as the relaxation energy of the lattice near a defect; a is the distance between equivalent positions of a defect; n is the number of equivalent directions of migration.

In the majority of solids the most mobile defects are interstitial atoms and ions. Typical values of the activation energy of their motion are 0.05–0.15 eV, so that they have a significant mobility at temperatures in excess of 10–50 K. The activation energy of vacancy motion is much higher and is usually 0.6–1.0 eV, so that their motion becomes significant only at temperatures in excess of 200–400 K.

There are two types of deviations from these general comments. Firstly, if migration of defects occurs as a result of their interaction with electron excitations, the effective activation energy of migration may be considerably less than in the absence of such interaction (this is known as radiation-stimulated or radiation-enhanced migration—see Sec. 3.3). Secondly, in some cases there is a high probability of the tunneling of defects across barriers separating equivalent positions and this is known as quantum diffusion. It is, in fact, important in the case of very light diffusing atoms at very low temperatures and has been discovered in a study of the recombination of hydrogen atoms in solid hydrogen at temperatures below 4 K (Ref. 26).

Migration of charged defects in an external electric field gives rise to ionic conduction of crystals, which is particularly important in the absence of equilibrium conduction electrons and valence bonds, i.e., in insulators.

In the course of migration in the lattice the spatially separated defects may approach one another or localized radiation or pre-irradiation defects, and can participate in defect–defect interactions. Such interactions destroy free elementary defects at a rate

$$-\frac{dN_d}{dt} = 4\pi N_d D_d \sum_i R_i N_i \tag{3.28}$$

[see Eq. (2.116)], where D_d is the diffusion coefficient of a mobile defect; N_d and N_i are the concentrations of a mobile defect and of the ith sink; R_i is the radius of capture of a mobile defect by the ith sink.

The quantity R_i is governed by an energy E_{bd} representing the interaction between defects and it can be estimated from the condition

$$E_{bd}(R_i) = E_m, \tag{3.29}$$

i.e., it can be deduced from the distance in which the energy of interaction between defects is equal to the energy of migration of a mobile defect.

Estimates are easiest to make if the interaction of defects is governed by the electrostatic forces. We then have

$$E_{bd} \approx \frac{Z_1 Z_2 e^2}{\varepsilon_0 r}, \tag{3.30}$$

where Z_1 and Z_2 are the charges of the interacting defects in units of e. If $Z_1 = Z_2 = 1$, it is convenient to use the expression $E_{bd} = E_B a_B / \varepsilon_0 r$. For the average values $\varepsilon_0 \approx 5$–6 we find that in the case of defects located in neighboring lattice sites and separated by $r \approx (6$–$7) a_B$ the interaction energies are $E_{bd} \approx 0.5$–1.0 eV and the corresponding distances are $R_i \approx (1$–$3) a$, in good agreement with the experimental results and more detailed calculations which give, for example, in alkali halide crystals values of about 0.5 eV for E_{bd} in the case of a pair with oppositely charged vacancies $(v_a^+ v_c^-)$ and 0.3–0.4 eV for a complex consisting of a doubly charged cation-replacing impurity and a cation vacancy $(M_c^{2+} v_c^-)$.

In the interaction of a charged defect with a neutral one characterized by the polarizability γ, the energy $E_{bd} \approx \gamma e^2 / 2\varepsilon_0 r^4$ can again represent only a few tenths of an electron volt if defects are located at neighboring lattice sites.

In the case of two identical defects the indistinguishability of their electronic states gives rise to an exchange interaction or, in other words, to covalent binding between them. Two F centers occupying the two nearest anion lattice sites in an alkali halide crystal have a binding energy of about 0.5–1.0 eV and form what is known as an F_2- (or M) center $(v_a^+ e)_2$. The interaction of two H centers may create an interstitial molecule $(X_i^0)_2$ or there may be even a considerable modifi-

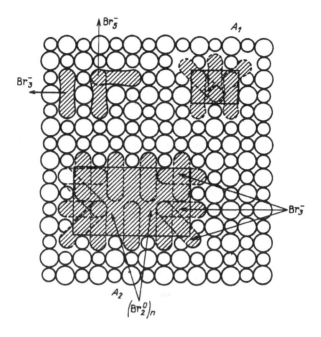

FIG. 3.18. Model of Br_3^- centers and of their coalescence producing small (Br_5^-, known as A_1) and large (A_2) bromine aggregates in an irradiated KBr crystal.[59]

cation of the lattice creating a quasimolecule X_3^- extending over three (two anion, one cation) lattice sites (Fig. 3.18). In the case of metals it is known that pairs of interstitial atoms (i_2) with a binding energy of about 0.5 eV (for example, 0.6 eV in copper) can arise, and this energy is also of the exchange origin.

In many cases the interaction between lattice defects via the surrounding elastic stress fields is of primary importance. This interaction is due to a change in the volume of a defect site and a reduction in the interatomic repulsion forces in the less densely packed regions of a crystal. Such regions are formed by vacancies or dislocation kinks, as well as by impurity atoms and ions. For example, the binding energy of interstitial atoms and ions with small-radius cation-replacing impurities in ionic crystals (Na^+ and Li^+ in KCl and KBr) is 0.5–1.0 eV. Defects generated in ionic crystals as a result of this interaction are identified by a subscript A (for example, an H_A center is an H center located alongside a singly charged cation-replacing impurity; an F_A center is an F center in a similar position).

Similar aggregates of intrinsic and impurity defects are also stable structure imperfections in semiconductors. For example, there are the well-known A centers (consisting of a vacancy near an interstitial oxygen impurity atom) and the E centers (consisting of a vacancy near a neutral donor such as a P atom) in silicon crystals, which may also contain interstitial atoms located close to oxygen impurity atoms. Divacancy centers are likely to appear in practically all solids.

The interaction of point defects with dislocations mediated by elastic stresses is described by the expression

$$E'_{bd} \approx \frac{YBa^2 \Delta a}{r},\qquad(3.31)$$

where Y is the Young modulus; B is the absolute value of the Burgers vector of a dislocation; r is the distance from the defect to the dislocation axis. If we assume that $Y \approx 10^{11}$ N/m^2 (which is typical of metals), $a \approx 0.3$ nm, and $\Delta a \approx \frac{1}{3}a$, we find that at a distance $r = B$ the interaction energy should be $E'_{bd} \approx 5$ eV.

The most cardinal consequences usually result from aggregation of radiation defects due to the interaction between defects of the same kind. The first stage of this process is the formation of defect pairs, some of which have already been mentioned. The interaction of single defects with pairs creates aggregates of three point defects [in the case of alkali halide crystals these are, for example, F_3 or R centers, whereas in photographic materials AgCl and AgBr there are $(Ag_i)_3$ coagulates known as precursor centers of photographic sensitivity]. If the migration energy of the defect pairs is less than their binding energy, then migration of such defects can play its role. Further coagulation of defects results in clusters of interstitial defects and vacancies and such clusters can be observed under an electron microscope.[56] There is a characteristic process of coagulation of anion vacancies (in the form of the F centers) in alkali halide crystals: it begins with the formation of large aggregates of the F centers (F_n or X centers, with $n \approx 10^2$–10^3), followed—because of a deficit of the halogen in such a region—by a phase transition creating metal particles (for example, Na in NaCl) known as the colloidal centers. Aggregation of interstitial halogen in such crystals is due to the formation of increasingly complex polyhalogens: two H centers create the X_2 or X_3^- quasimolecule, four H centers produce X_5^-, and so on until the precipitate is so large that it can be regarded as a gas bubble consisting of $(X_2^0)_n$ (Fig. 3.18). Formation of large silver aggregates (photosensitivity centers) in silver halides involves successive capture of holes and Ag_i^+ ions by precursor centers. A common property of large radiation-defect coagulates is their high stability compared with smaller aggregates.

It is important to note that aggregation of like radiation defects may occur even without migration. In view of the spatially random creation of v–i pairs and preferential recombination of the nearest v and i from different pairs, we have a situation when like defects "prefer" to form clusters, because the probability of survival of one defect in a cluster is higher than in a random distribution.[27]

If the energy stored in radiation defects exceeds a certain critical value, an irradiated substance may exhibit a phase transition from a crystal to an amorphous substance. By way of example, Fig. 3.19 shows the K_α x-ray emission spectra of diamond irradiated with fast neutrons. These spectra reflect the structure of the valence band (Sec. 2.4). When the neutron dose (fluence) reaches $(4.6$–$9) \times 10^{20}$ cm^{-2}, there is an abrupt change in the spectra indicating a phase transition from diamond to graphite.

It is interesting to note that defect clusters sometimes become ordered forming

S, rel. units

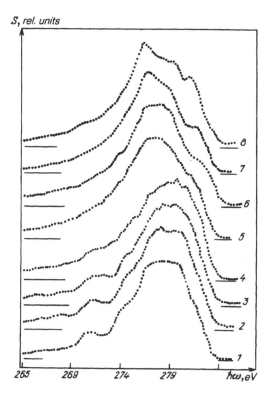

FIG. 3.19. X-ray *K* emission spectra of diamond before (1) and after (2) irradiation with fast neutrons in various doses (10^{19} cm^{-2}): (2) 1.5; (3) 14; (4) 46; (5) 97; (6) 104. Curves (7) and (8) represent the corresponding spectra of soot (7) and graphite (8).[34]

what is known as a superlattice (see, for example, Ref. 82). This phenomenon belongs to a wide range of self-organization processes which are the subject of synergetics.

Impurity atoms introduced in solids in the course of irradiation represent a special class of radiation defects. For example, irradiation with protons or α particles creates atoms of hydrogen and helium in the irradiated substance and these coagulate to form bubbles of the relevant gas. These and other impurity atoms arise as a result of nuclear reactions, which are particularly important when matter is irradiated with neutrons (Sec. 3.1). Bombardment of KCl crystals with C^+ and N^+ ions is reported to create molecular CN^- centers.[111]

One of the most common procedures in modern semiconductor technology is ion implantation doping, which represents introduction of the required impurities into a semiconductor by irradiation with impurity ions. Since high-energy heavy ions cause ionization and create structure defects in crystals, this doping method is frequently accompanied by undesirable effects. Nevertheless, it is frequently a convenient way of creating the necessary spatial distribution of impurities in a crystal, particularly when a doped surface layer with special electric properties is needed (Fig. 3.20). The channeling effect (see Sec. 3.1) can sometimes be used to control the thickness and properties of such surface layers. The undesirable crystal struc-

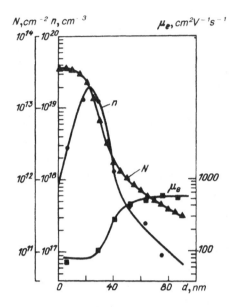

FIG. 3.20. Distribution with depth in Si of impurities (N), free carriers (n), and carrier mobility (μ) after a dose of 2×10^{14} cm^{-2} of B$^+$ ions ($E_0 = 40$ keV, annealing at 1000 K).[108]

ture defects created by ion implantation doping can be largely removed by subsequent annealing. Laser annealing is used increasingly for this purpose (this is discussed later).

Modifications of the electron and nuclear subsystems of a crystal as a result of irradiation with ionizing particles or photons disturbs their translation symmetry and, therefore, changes the majority of the physical properties. These changes determine the practical applications of solids exposed to radiation and can provide information on the nature of radiation defects.

We shall first consider the changes in the mechanical properties such as the size and the strength. These changes are due to the fact that defects are surrounded by elastic stress fields and that lattice relaxation near defects alters the volume of defective lattice sites.

In the case of the Schottky defects the atoms removed from regular lattice sites are located on the surface of a crystal. A change in the volume of a crystal ΔV_S consists of changes in the volume due to these atoms $n_S v$ (where n_S is the number of the Schottky defects and v is the volume of one lattice site) and changes in the volumes of the defect sites $n_S \Delta v_v$ (Δv_v is the change in the volume due to one vacancy), so that $\Delta V_S = (v + \Delta v_v)n_S$. If the total number of atoms in the lattice is N, then the total volume of the crystal is Nv. Therefore, the relative change in the total volume is

$$\left(\frac{\Delta V}{V}\right)_S = \left(1 + \frac{\Delta v_v}{v}\right)\frac{n_S}{N}, \qquad (3.32)$$

and the relative change in its lattice constant is

$$\left(\frac{\Delta a}{a}\right)_S = \frac{1}{3}\frac{\Delta v_v}{v}\frac{n_S}{N}. \tag{3.33}$$

In the case of the Frenkel defects the atoms removed from the sites remain in a crystal at interstitial positions. If the change in the volume of a crystal due to one interstitial atom is Δv_i, the change in the volume of the crystal is

$$\left(\frac{\Delta V}{V}\right)_F = \frac{\Delta v_i + \Delta v_v}{v}\frac{n_F}{N} \tag{3.34}$$

and the change in the lattice constant is

$$\left(\frac{\Delta a}{a}\right)_F = \frac{1}{3}\frac{\Delta v_i + \Delta v_v}{v}\frac{n_F}{N}. \tag{3.35}$$

If independent measurements of ΔV (for example, by determination of the macroscopic dimensions of a crystal or its density) and Δa (for example, by x-ray structure methods) are made, it is possible to determine the nature of the newly created defects. In the case of the Schottky defects, it follows from Eqs. (3.32) and (3.36) that

$$\frac{\Delta V}{3V} = \left(1 + \frac{v}{\Delta v_v}\right)\frac{\Delta a}{a}, \tag{3.36}$$

whereas for the Frenkel defects, we obtain the following expression from Eqs. (3.34) and (3.35):

$$\frac{\Delta V}{3V} = \frac{\Delta a}{a}. \tag{3.37}$$

This method has been used, for example, to show that irradiation of alkali halide crystals with x rays creates mainly the Frenkel defects, whereas additive coloration of these crystals produces the Schottky defects (Fig. 3.21).

The interaction of point defects with dislocations is responsible for the accumulation of defects near dislocations and this results in pinning of dislocations and increases the parameters of a crystal which represent its hardness, such as the Young modulus, yield stress, microhardness, etc. (Fig. 3.22). The main effect is due to interstitial atoms, which correspond to high values of Δa.

Radiation segregation is a similar effect: it results in spatial separation of different components of metal alloys and occurs when these alloys are irradiated in the presence of mechanical stresses or other inhomogeneities; it is the result of different directions of migration of defects of different size.

At high defect concentrations the leading role is played by large-scale disturbances of the chemical binding and by phase transitions in the irradiated substance. They result in swelling, sputtering, and complete fracture of crystals. For example, irradiation of alkali halide crystals with large electron or x-ray doses results in the periodic emission of large numbers of anions and cations.[53] Irradiation of solids with ultrahigh-power (in excess of 10^{10} W/cm^2) electron or laser radiation pulses causes decomposition.[13,40] A special case of damage to solids is the formation of light gas (hydrogen and helium) atoms. When the concentrations of these atoms

Nuclear subsystem

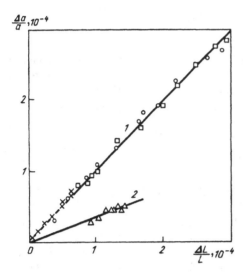

FIG. 3.21. Relationship between the relative changes in the lattice constant a and the length L of a KCl crystal irradiated with x rays (1) or heated in a potassium vapor (2), based on data from Ref. 63.

are high, they congregate to form gas bubbles which cause macroscopic swelling and bending of irradiated samples.

Radiation-induced changes in the optical properties play an important role in the physics of insulators. These changes are manifested by the appearance of new

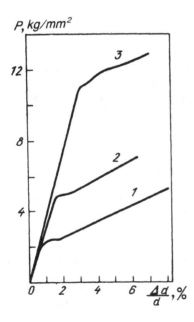

FIG. 3.22. Dependences of the relative compression on the pressure applied to an MgO crystal in various states: (1) unirradiated; (2) irradiated with an electron dose of 1.2×10^{18} cm^{-2} ($E_0 = 1.8$ MeV); (3) irradiated with a neutron dose of 8.3×10^{16} cm^{-2} ($E_0 \gtrsim 1$ MeV). Data taken from Ref. 110.

absorption and luminescence bands in the transparency ranges of the irradiated crystals and are due to transitions between electron states of defective parts of the lattice.

Using Eqs. (2.46), (2.23), and (2.25) for the absorption coefficient associated with such transitions (between two discrete states m and k), we find that

$$\mu_{mk} = N \frac{4\pi^2 e^2 \hbar \omega_{mk}}{3c\hbar} |r_{mk}|^2 = N f_{mk}(\omega) \frac{2\pi^2 e^2 \hbar}{mc}. \tag{3.38}$$

If a $k \to m$ transition is represented by an absorption band rather than a line, integration over this band has to be carried out. We then have

$$N f_{mk} = \frac{mc}{2\pi^2 e^2 \hbar} \int \mu_{mk}(\omega) d(\hbar\omega),$$

where $f_{mk} = \int f_{mk}(\omega) d(\hbar\omega)$ is the total oscillator strength of a transition between electronic states m and k. In the case of a Lorentzian band profile, we have

$$\int \mu_{mk}(\omega) d(\hbar\omega) = \pi \mu_{max} \frac{\Delta W_{1/2}}{2},$$

where μ_{max} is the absorption coefficient at the band maximum and $\Delta W_{1/2}$ is the width of the band at half-height.

We must also allow for the fact that the absorbing centers are not in vacuum but in a medium with a refractive index n, so that the substitutions $c \to c/n$ and $A_0 \to 3A_0/(n^2 + 2)$ have to be made. After these substitutions, we obtain

$$N f_{mk} = \frac{9mcn}{4\pi e^2 \hbar (n^2 + 2)^2} \mu_{max} \Delta W_{1/2}, \tag{3.39}$$

and inclusion of numerical values in the constants modifies the above expression to

$$N(\text{cm}^{-3}) f_{mk} = (10^{16} - 10^{17}) \mu_{max}(\text{cm}^{-1}) \Delta W_{1/2}(\text{eV}).$$

The best known radiation-induced bands are the F absorption and luminescence bands in the spectra of ionic crystals, due to optical transitions in the F centers. Bearing in mind that an F center (an electron in the field of an anion vacancy) is phenomenologically similar to a hydrogen atom, a comparison with the optical properties of the latter is the simplest approach that can be used in discussing the optical properties of the F centers. We can therefore expect the appearance of an absorption band due to a dipole-allowed $1s \to 2p$ transition in an F center at photon energies $\hbar\omega_F \approx e^2/2\varepsilon_0 a$, i.e., in the visible part of the spectrum. Such a band is indeed observed in the spectra of irradiated alkali halide crystals (Fig. 3.23), the result of which is a characteristic coloration of these crystals. The energy at the maximum of the F absorption band $\hbar\omega_F$ changes from crystal to crystal in accordance with the Molwo–Ivy rule $\hbar\omega_F a^2 \approx \text{const}$, where a is the lattice constant, i.e., as a is increased, this absorption band shifts toward lower photon energies. The oscillator strength of an optical transition in the F absorption band is 0.5–0.8.

An excited F center may drop to the ground state emitting a photon. Since in the course of relaxation of an excited F center its energy decreases considerably and the $2p$ and $2s$ states become mixed, the Stokes shifts between the F absorption and

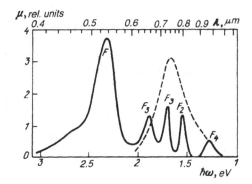

FIG. 3.23. Absorption spectra of the F_n (continuous curve) and colloidal (dashed curve) centers in a KCl crystal.[44]

luminescence bands are large (the F luminescence band is usually located at about 1 eV) and the lifetime of a relaxed excited F center is relatively long (≈ 1 μs).

The absorption and luminescence bands of the F_A centers are somewhat shifted toward lower energies compared with the corresponding bands of the F centers. In view of the influence of the adjacent impurity the F_A centers are much more stable than the F centers, which makes it possible to use crystals with the former centers in lasers. Such color-center lasers emit in the spectral range from 1 to 3 μm, which is relatively inaccessible to lasers of other types.

Characteristic absorption bands are exhibited also by the aggregate F centers in ionic crystals. For example, optical transitions in the F_2 centers can be considered on the basis of the model of the H_2 molecule. The main absorption band due to the $^1\Sigma_g^+ \rightarrow {}^1\Sigma_u^+$ transition is exhibited by KCl crystals at 1.5 eV and the corresponding luminescence band at 1.2 eV. The $^1\Sigma_g^+ \rightarrow {}^1\Pi_u$ transition at higher energies lies within the F absorption band (2.1 eV). In the range 1.2–2.0 eV there are also absorption bands of the F_3 and F_4 centers and also of the colloidal centers (Fig. 3.23). The band of the colloidal centers can be described satisfactorily as the result of the scattering of light by microscopic metal particles.

Quasimolecular local vibrations in $F_{2(3)}$ centers are responsible for the fact that the low-temperature profiles of their absorption and luminescence bands are typical of the luminescence centers with low Stokes losses and they include a zero-phonon line.[84]

The absorption bands of halogen defects in alkali halide crystals are located in the ultraviolet part of the spectrum. In their relaxed state the primary halogen defects are X_2^- quasimolecules located either at two anion lattice sites (self-localized holes or V_K centers—see Sec. 2.5) or at one site (interstitial halogen atoms or H centers). Their optical properties can be considered on the basis of a scheme of states of free X_2^- ions. In accordance with this scheme, the X_2^- centers have one absorption band in the ultraviolet part of the spectrum ($^2\Sigma_u^+ \rightarrow {}^2\Sigma_g^+$ transition at 3.4–3.7 eV in KCl) and another in the visible range ($^2\Sigma_u^+ \rightarrow {}^2\Pi_g$ transition at 1.6–2.4 eV in KCl). Very similar optical properties are exhibited also by the V_{KA} and H_A centers. The absorption band of the X_3^- centers ($^1\Sigma_g \rightarrow {}^1\Sigma_u$ transition) in the ultraviolet (at about 4.5 eV for KBr), is shown in Fig. 3.24 (Refs. 38 and 96).

The absorption of photons by the host substance is also perturbed in the vicinity

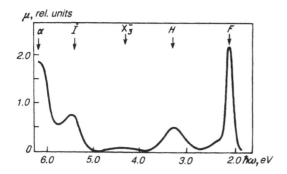

FIG. 3.24. Absorption spectra of a KBr crystal after irradiation with x rays at 5 K (Ref. 38).

of lattice defects. Since usually either an electron or a hole is bound to a defect, the exciton energy near defects is usually less than in an ideal lattice. This results in a low-energy shift of the fundamental absorption edge of crystals with defects. There are clear bands in the fundamental absorption region of alkali halide crystals and these are known as the α, β, and γ bands: they are due to the appearance of excitons near an anion vacancy, an F center, and possibly an interstitial halogen ion, respectively.

When a crystal contains impurity centers capable of capturing electrons or holes, irradiation gives rise to absorption bands associated with the recharge of these centers. For example, the spectra of KCl:Ag crystals have bands associated with the Ag^0 and Ag^{2+} centers, which are formed as a result of the capture of electrons and holes by the Ag^+ centers, as well as a band due to the Ag_a^- centers, due to the displacement of Ag^+ from a cation site to an anion one.[44]

Irradiation of semiconductors also alters their optical properties. However, since the band gaps of semiconductors are relatively narrow, the absorption bands are located in the infrared. For example, the absorption spectrum of irradiated silicon has a number of poorly resolved bands at 2–30 μm. After irradiation with large doses this element acquires a smoky color associated with an amorphous phase and due to changes in the function representing the density of the band states. The absorption spectrum of GaAs has a band at 1.0 eV and this is clearly an analog of the α band in the spectra of alkali halide crystals.[11] The absorption spectrum of irradiated n-type InSb has a band at 0.093 eV which is probably associated with a transition of an electron from an indium vacancy to the conduction band (Fig. 3.25).

In studies of optical transitions of defects in semiconductors, the efficiency of absorption of incident light is frequently measured not on the basis of the intensity of the transmitted light (as in the case of ionic crystals), but in terms of the efficiency of liberation of carriers from defects and the appearance of an electric current, i.e., determination of the absorption spectra is replaced by recording of the photoconductivity spectra. Detailed information has been obtained in this way on the spectrum of radiation-defect levels of many semiconductors.

A change in the optical density of solids due to irradiation with ionizing particles of photons is of twofold practical importance. On the one hand, it damages optical devices and components used in the presence of strong radiation (blackening of the

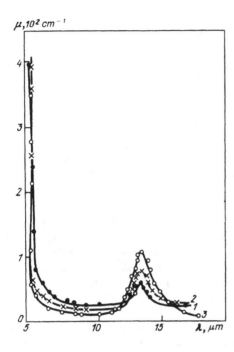

FIG. 3.25. Absorption spectra of n-type InSb recorded at 100 K after irradiation with various doses of fast neutrons (10^{16} cm^{-2}): (1) 8.1; (2) 50; (3) 300. Data taken from Ref. 12.

optical components). On the other hand, these changes can be used to store information carried by radiations. Alkali halide crystals with color centers provide a good example of a medium suitable for recording volume holograms. More complex compounds in the form of sodalites, in which very stable but still easily removable F centers can be formed, have proved to be good materials for storing visually recorded data (for example, printed text) by an electron beam.[21] Finally, silver salts are the most common photographic materials.

Irradiation alters also the magnetic properties of substances. As a rule, the radiation-induced changes in the structure weaken the dominant type of magnetism of a given substance. In the case of ferromagnets there is a reduction in the Curie temperature and in the Curie constant.[112] Irradiated paramagnets may exhibit diamagnetic or ferromagnetic phases (the latter have been observed, for example, on irradiation of stainless steel with He$^+$ ions of 40 keV energy in doses in excess of 8×10^{17} cm^{-2} applied at 500 K—such phases are the result of segregation of nickel atoms[90]). However, in the case of alloys the radiation-induced segregation may produce the opposite effects. For example, irradiation of an Fe–Ni alloy with high-energy electrons can increase the Curie temperature.[113]

In the case of diamagnetic materials the most typical change is the appearance of paramagnetic centers. In principle, these are all the defects with unpaired electrons. In the presence of an external magnetic field the spin-degenerate energy levels of such defects are split by an amount $\Delta E = g\beta H$ (where g is the spectroscopic g factor, β is the Bohr magneton, and H is the magnetic field intensity), so that optical transitions between the sublevels can now take place. The fine and hyperfine structures of the transitions, which are due to the interaction of a spin of a given electron with the spins of other electrons and the surrounding nuclei, are very informative sources of our knowledge of the structure of the centers and their environment. The method employed is known as electron spin or paramagnetic resonance (ESR) and it is used widely in studies of radiation effects in insulators and semiconductors.

The interaction of radiations with solids may alter the structure of their surface. The specific effects of irradiation of the surfaces, compared with the bulk, are primarily due to two circumstances. Firstly, some defect formation processes may appear effectively on the surface even if they are hindered in the bulk by omnidirectional close packing. They include, for example, the mechanisms associated with the radiation-induced electrostatic instability (see Sec. 3.3). Secondly, crowdions and channeled atoms, initially created inside a crystal, appear on the surface (Sec. 3.2). The net result is the disturbance of the surface structure which may differ somewhat from the disturbance caused in the interior of a solid. For example, irradiation of glasses and oxides results in depletion of the surface layer in respect of some elements and a corresponding enrichment with others (for example, sodium silicate glasses are depleted of sodium and oxygen[129] and the surfaces of ionic crystals exhibit particles or even layers of a pure metal as a result of preferential escape of the anion component[114]). Carriers localized at capture (trapping) centers located near the surface of a crystal accelerate the adsorption of oxygen molecules on the surface.[6] On the other hand, electron excitations created on the surface or migrating to the surface from the interior may result in desorption of adsorbed foreign atoms and molecules (this effect is used in radiation cleaning of surfaces).

An investigation of the spectrum and number of atoms and ions escaping from solids can provide information on the composition of surface layers of the irradiated samples and of the processes which occur there. A popular method used for this purpose is secondary-ion mass spectrometry (SIMS) in which a sample is irradiated with fast ions and the products are analyzed by mass spectrometry.

The accumulation of stable radiation defects and manifestation of the associated changes in the properties of crystals are possible only under conditions such that both primary defects (a vacancy and an interstitial atom) are localized either in the regular lattice or as a result of the interdefect interaction. In the opposite case we can expect recombination of i with v and restoration of the regular lattice. In the case of some crystals with a loose structure (for example, In_2Te_3) the conditions are unfavorable for the localization of interstitial atoms and stable radiation defects do not appear at all.[22] In the case of those crystals in which primary defects become localized at low temperatures, an increase in temperature increases their mobility and this gives rise to their recombination with immobile partners. A procedure resulting in liquidation of defects by heating of crystals is known as the annealing of defects.

Bearing in mind the activation energies of motion and the defect binding energies mentioned above, we can describe in general the annealing of defects as follows.

At temperatures 5–50 K in most solids we can expect the onset of rapid motion of interstitial ions and atoms and the result is that some of them recombine with vacancies, while others become localized because of the interaction with other radiation or preradiation defects. The lowest-temperature stages of annealing are due to the recombination of what are known as close defect pairs, i.e., pairs sepa-

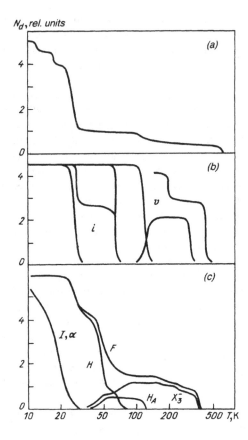

N_d, rel. units

(a)

(b)

v

i

(c)

I, α

F

H

H_A X_3^-

10   20   50   100   200   500 T, K

FIG. 3.26. Annealing of radiation defects in a metal (platinum, a, data taken from Ref. 130), in a semiconductor (germanium, b, Ref. 74), and in an ionic crystal (KBr, c, Refs. 78 and 96).

rated by short internal distances, which are mainly genetically related pairs (those created simultaneously at the same act). At temperatures 50–200 K we can expect liberation of interstitial ions and atoms from various traps, which results in further liquidation of some of them and a more stable localization of the others (including formation of large aggregates). At temperatures 200–400 K the remaining single vacancies and their small aggregates begin to move rapidly; the vacancies then partly recombine with interstitial atoms which are components of stable complexes and partly form larger and more stable vacancy complexes. Finally, intense ionic processes resulting in destruction even of relatively stable complexes of interstitial atoms and vacancies begin at 400–600 K. At these temperatures we can expect the last stages of annealing which completes liquidation of practically all the radiation defects. Only very strongly damaged materials exhibit at these temperatures coagulation of some of the microdefects giving rise to macrodefects, such as voids (negative crystals).

Examples of the various stages of annealing of defects in an ionic crystal of KBr, a semiconductor Ge, and a metal Pt are demonstrated in Fig. 3.26.

The recent developments in laser technology have made laser annealing of defects a popular treatment. High-intensity laser radiation beams provide convenient

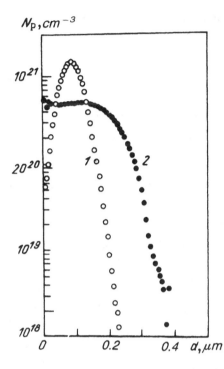

FIG. 3.27. Distribution of P atoms in depth in Si before (1) and after (2) laser annealing. The P^+ ions $(1.16 \times 10^{16}$ cm$^{-2})$ were introduced by implantation $(E_0 = 80$ keV). Data taken from Ref. 135.

and rapid means for controlling the temperature of solids in a wide range and with specified geometry (i.e., this method is spatially selective). An increase in temperature and the propagation of heat in an irradiated object is described by Eq. (3.16), where Q now represents the laser radiation energy absorbed in an object. If, for example, a laser pulse of 1.5 J/cm^2 energy is absorbed in a silicon layer 10 μm thick, then each atom receives on the average an energy of 5 eV, which is four times greater than the band gap E_g and is an order of magnitude higher than the heat of melting. Figure 3.27 gives an example of the effects of laser annealing of silicon crystals doped with P atoms by the method of ion implantation.

Interference between two laser beams on the surface of an annealed sample can produce a structure with a period close to the wavelength of laser radiation. Such interference laser annealing provides new opportunities for the fabrication of semiconductor microelectronic devices.

3.5. Defects in molecular systems

The primary result of the fragmentation of molecules, which corresponds to the formation of the primary defects in crystals, is the destruction of molecules as such and formation of free radicals (see Sec. 3.3).

Free radicals are molecules or their fragments with an unpaired electron, which is usually delocalized and distributed over the whole molecule. They are chemically active and tend to form bonds with surrounding molecules. In the case of condensed systems one of the most probable reactions of the resultant radicals is their recombination which restores the original molecule, representing an analog of the interstitial–vacancy $(i + v)$ recombination in crystals (see Sec. 3.4), which is frequently called the cell effect.

The general pattern of the products of secondary reactions involving such primary radicals can be very complex. By way of example, we shall now list the more important secondary reactions following radiolysis of water:

$$H_2O^* + H_2O^* \rightarrow H_2O_2 + H_2, \quad H_2O^+ + H_2O \rightarrow H_3O^+ + OH,$$

$$H^+ + H_2O \rightarrow H_3O^+, \quad O^+ + H_2O \rightarrow H_2O^+ + O,$$

$$O^- + H_2O \rightarrow OH + OH^-, \quad H^- + H_2O \rightarrow H_2 + OH^-.$$

A chain of primary (see Sec. 3.3) and secondary processes creates stable products of radiolysis of water:

$$H, \ OH, \ e_{aq}^-, \ H_2, \ H_2 O_2, \ H_3 O^+.$$

Specific examples of radiation-chemical reactions are isomerization (see Sec. 3.3), dimerization, polymerization, and excimerization. Isomerization is the reaction that alters the geometric structure of a molecule without changing its chemical composition. If the change in the structure involves rotation of one part of a molecule relative to another, we are dealing with tautomerization. Dimerization is formation of one molecule from two molecules present initially. A special case of dimerization is when the two original molecules are identical.

In some cases the reactions between excited molecules may create more complex molecules which are stable only in the excited state and dissociate when the excitation decays, for example, by the emission of a photon. These molecules are called excimers. Examples of excimers are rare-gas molecules such as $(Ar_2)^*$, and so on, as well as compounds of rare-gas atoms with halogen atoms $(ArF)^*$, etc. The latter are used in what are known as excimer lasers.

An important type of reaction exhibited by irradiated molecules is polymerization. This process consists of the joining of small molecules into large chains. The role of radiation consist in the excitation or ionization of the original molecules, which greatly accelerates the process. On the other hand, radiation can also cause breaking of bonds of the existing polymer molecules and can thus have a destructive influence on polymers. Therefore, polymers can be divided into two groups from the point of view of the interaction with radiation: those which are predominantly joined together and those which are mainly destroyed.

By way of example, we shall consider the effects of radiation on polyethylene $(CH_2)_n$. The primary reaction is the breaking of the C—H bonds and the formation of unsaturated groups:

$$-CH_2-CH_2-CH_2- \Big\langle \begin{array}{l} {}^{\nearrow}-CH_2-\dot{C}H-CH_2- + H \\ {}^{\searrow}-CH_2-CH=CH-CH_2- + H_2 \end{array}$$

The interaction between adjacent molecules with unsaturated bonds results in cross-linking and the interaction with oxygen molecules induces oxidation. Both processes alter the macroscopic properties of polyethylene. It should be pointed out that irradiated polyethylene has the same electrically insulating properties as unirradiated polyethylene but they are retained to higher temperatures; moreover, the

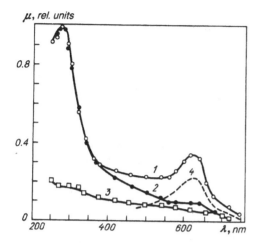

μ, *rel. units*

FIG. 3.28. Absorption spectra of crystalline ice at 77 K: (1) after irradiation with a γ-ray dose of 4×10^{20} eV/g; (2) after subsequent illumination with visible light; (3) after annealing at 123 K; (4) spectrum of e_{aq}^- obtained by subtracting spectrum 2 from spectrum 1 (Ref. 46).

irradiated product has a size memory: when such polyethylene is extended (as to increase its linear dimensions by 50–100 %) and then heated, it recovers its initial dimensions.

Among the physical methods used in investigations of radiation-chemical processes the most informative are optical and magnetic techniques, mainly determination of the absorption, luminescence, and ESR spectra, and in the case of gaseous phases one can also use the method of mass spectroscopy.

Some products of radiolysis of water give rise to characteristic bands in the absorption spectra: e_{aq}^- corresponds to a band at 600–800 nm, OH to that at 260–300 nm, O_2^- to that at 240–260 nm, etc. Figure 3.28 shows the absorption spectrum of crystalline ice irradiated with γ rays, which is dominated by two bands; one at 280 nm, due to the OH radicals, and the other at 620 nm due to e_{aq}^-.

Irradiation of an aqueous solution of ferrous sulfate oxidizes the divalent iron to the trivalent state. Among the many possible processes the main role is played by the reaction $Fe^{2+} + H_2O_2 \rightarrow Fe^{3+} + OH + OH^-$. The Fe^{3+} ions have absorption bands at 305 and 224 nm when in solution. The intensities of these bands can be used to determine the concentration of the Fe^{3+} ions and, consequently, the radiation energy or dose absorbed in the solution (this is used in the ferrous sulfate or Fricke dosimeter).

The main primary product of the radiolysis of nitrates containing the NO_3^- molecular group is the molecular ion NO_2^- formed as a result of the primary $NO_3^- \rightarrow NO_2^- + O$ reaction or the secondary $NO_3^- + O \rightarrow NO_2^- + O_2$ reaction. It gives rise to an absorption band at 350 nm and is manifested also in the ESR spectra.

One of the typical results of the effects of radiation on organic molecules is the formation of closely spaced (frequently within the same molecule) radical pairs. An example is the reaction

$$R-N{=}N-R \rightarrow \dot{R}\,N_2\,\dot{R}$$

where $R \equiv (CH_3)_2CCN$, which occurs in liquid azobisisobutyronitrile when the average distance between the nearest $\dot{R}$ radicals is 0.56 nm. Another example is the result of radiolysis of solid methane at 4 K, representing the formation of pairs of $\dot{H} \cdots \dot{C}H_3$ radicals with an average distance 0.67 nm (Ref. 105).

Radiation-chemical processes have probably played an important role in the formation of the very first organic molecules and, therefore, of life on Earth. It is assumed that during the first stages of the evolution of the terrestrial atmosphere it consisted of molecules of hydrogen (H_2), water (H_2O), carbon dioxide (CO_2), methane (CH_4), and ammonia (NH_3). Ultraviolet radiation, cosmic rays, and electric discharges broke down the C—H, H—O, and H—H bonds. After recombination of the products new stable molecules were formed including the hydrocyanic acid H—C≡N, formic acid

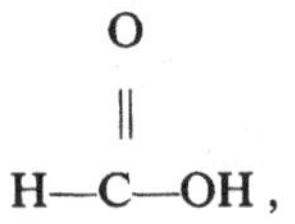

$$\begin{array}{c} O \\ \| \\ H-C-OH \, , \end{array}$$

and formaldehyde

$$\begin{array}{c} H \\ | \\ H-C{=}O \, . \end{array}$$

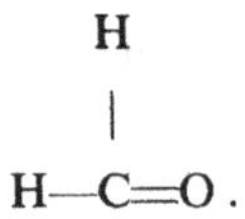

The action of ionizing radiations on these molecules produced even more complex molecules, including amino acids, for example, allicin,

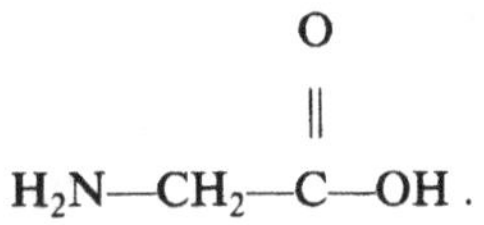

$$\begin{array}{c} O \\ \| \\ H_2N-CH_2-C-OH \, . \end{array}$$

alanine,

$$\begin{array}{c} O \\ \| \\ CH_3-CH-C-OH \, , \\ | \\ NH_2 \end{array}$$

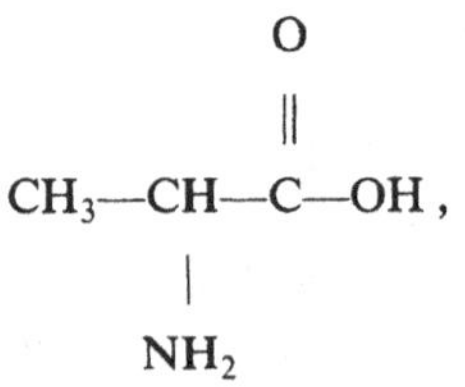

and others.

The ability to form complex molecules by irradiation of a mixture of simple molecules in closed containers has been demonstrated experimentally: irradiation of aqueous solutions and suspensions of carbonates with x rays and γ rays generates oxalic acid and traces also of formic and glycolic acids, and formaldehyde.[32]

Elementary physical processes due to irradiation of macromolecules are in principle similar to those discussed above, but the greater dimensions give rise to certain specific features in the behavior of macromolecules:

● some macromolecules (enzymes, DNA, etc.) have unique biological properties which are lost as a result of irradiation;

● in the majority of cases the energy received by a macromolecule from radiation is released as a form of damage to the structure, not at the place where it is received, but after intramolecular or intermolecular migration (see Sec. 2.5) at some other point of the molecules;

● macromolecules which are in some medium may be damaged not only as a result of the direct interaction with radiation, but also because of the interaction of radiation of the medium itself (indirect radiation effects) causing creation of small free radicals which in many liquid solvents can diffuse over a distance $(D\tau)^{1/2}$ of the order of 3 nm $(D \approx 10^{-5} \text{ cm}^2/\text{s}, \ \tau \approx 10^{-8} \text{ s})$;

● the first stage of disturbance of the functional properties of macromolecules is the formation of latent defects, which by themselves do not affect these properties, but are developed or repaired as a result of subsequent effects of various radiation or nonradiation factors on the molecule.

One of the most important phenomena in radiation biology is the oxygen effect: the damaging action of radiations is greatly enhanced in the presence of oxygen and water, compared with irradiation under anaerobic conditions. It is important to stress that oxygen can frequently influence the effects after irradiation by acting on latent damage or molecules. Some of the damage may occur a very long time after the end of irradiation and some may occur only in the presence of oxygen during irradiation. This is related to the various values of the lifetime of the various types of latent damage.

The mechanism of the oxygen effect reduces to the interaction of the O_2 molecules with radicals $\dot{R}$ created by irradiation. We shall give here some examples:

$$\dot{R} + O_2 \rightarrow \dot{R}O_2, \quad \dot{R}O_2 + RH \rightarrow RO_2H + \dot{R}, \quad \dot{R}O_2 + O_2 \rightarrow \dot{R}O + O_3.$$

The indirect effects of radiation and latent damage provide a chance of some control in the radiation damage to macromolecules by a change in the composition of the surrounding medium during and after irradiation, ensuring opportunities for protection and repair. The mechanisms of these processes can be divided into two classes: the capture of active molecules by protective molecules (which is known as competition) and liquidation of latent damage (restitution). For example, irradiation of RH molecules can create radicals which represent latent damage:

$$RH + \dot{O}H \rightarrow \dot{R} + H_2O.$$

Such latent damage may be realized as a result of interaction with oxygen molecules:

$$\dot{R} + O_2 \rightarrow \dot{R}O_2,$$

and it may be repaired by the interaction with SH radicals:

$$\dot{R} + SH \rightarrow RH + \dot{S}.$$

It therefore follows that the addition to the irradiated medium of molecules containing the SH group can repair the radiation damage of the RH-type molecules.

In the case of such indirect action of radiation the temperature during irradiation has a strong influence on the results: the higher the temperature the stronger the effects of radiation. The effects are due to the acceleration of diffusion of free radicals on increase in temperature.

When macromolecules are irradiated, one frequently encounters situations when the final damage or inactivation of a molecule occurs as a result of realization of latent damage by second interaction with radiation. Therefore, the effects of irradiation are usually enhanced on increase in the linear energy losses $|dE/dx|$ of the incident radiation in matter. This property of radiation can be represented by the relative biological efficiency defined as the γ-radiation dose divided by the dose of a given radiation producing the same effect.

The maximum relative biological efficiency (up to 20) is exhibited by heavy charged particles and neutrons (the action of the latter reduces mainly to the effects of secondary charged particles). The specific value of the relative biological efficiency depends on the object being irradiated: the more complex the object, the more important the role of any latent damage and the stronger the dependence of the relative biological efficiency on $|dE/dx|$.

If damage requires a certain limited number of the interactions of the molecule with radiation, then in the case of large linear losses the relative biological efficiency again decreases because an increase in the number of interactions makes some of them redundant.

An increase in the value of $|dE/dx|$ reduces the oxygen effect (because the latent damage is realized by the radiation itself and oxygen is no longer required) and weakens the repairing influence of other molecules.

We shall finally consider the interaction of radiation with the molecules of DNA. These molecules represent a biopolymer consisting usually of two chains twisted to form a double helix (Fig. 3.29). Each of these chains consists of molecules of four bases (adenine, thymine, guanine, and cytosine) and also of molecules of deoxyribose (sugar) and phosphate groups. These bases are bound to the sugars by a relatively strong N-glycoside bond and to one another by weak hydrogen bonds; only the adenine–thymine and guanine–cytosine bonds are possible. The total length of the molecule represents 10^4–10^9 pairs of bases.

The main forms of the radiation damage to DNA are shown schematically in Fig. 3.30. One of the most frequent of them, particularly in the case of irradiation with ultraviolet photons, is the formation of dimers of the bases, particularly of thymine which occurs in accordance with the scheme shown in Fig. 3.31.

Irradiation of DNA also alters the composition of the bases and breaks the nucleotide bonds (it causes depolymerization). This affects particularly the P—O, C—C, and C—O bonds as well as the hydrogen bonds. Another type of radiation damage is cross-linking, i.e., false bonds between the chains. An indirect effect of ionizing radiations on these molecules generates primarily the radicals OH, H, and e_{aq}^-.

The minimum energy necessary to damage a DNA molecule is 4–5 eV. An absorption band with its long-wavelength part clearly due to the transitions in thymine is located at corresponding wavelengths (Fig. 3.32). The spectra of photoformation of chromosome aberrations of the cells of the Chinese hamster and the photodestruction of the *E. coli* strain are well correlated with this band.

Selective photofragmentation of macromolecules provide a major research tool and would be of great practical importance. However, the attainment of such selectivity is a very difficult task, because the absorption bands of different macromolecules and different parts of one macromolecule overlap strongly. It would be

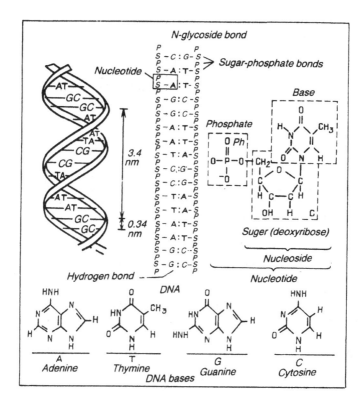

FIG. 3.29. Schematic representation of the DNA structure.[42]

necessary to develop special procedures, an example of which is what is known as two-photon affine modification.[9] Use is made of the circumstance that some dyes become bound to macromolecules and, consequently, transfer energy selectively. For example, molecules of the dye 8-methoxypsoralen (methoxsalen) form complexes with DNA, but do not become bound to RNA or proteins. If a solution containing all these types of molecules is irradiated with nitrogen laser radiation (337 nm), then this radiation is absorbed only by the dye molecules. After successive absorption of two photons such a molecule is in an excited state and its energy

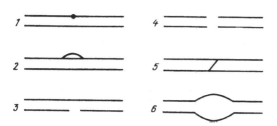

FIG. 3.30. Typical structure defects in the double helix of the DNA: (1) damage to the bases; (2) formation of dimers; (3) single rupture; (4) double rupture; (5) double link; (6) longitudinal rupture of hydrogen bonds.

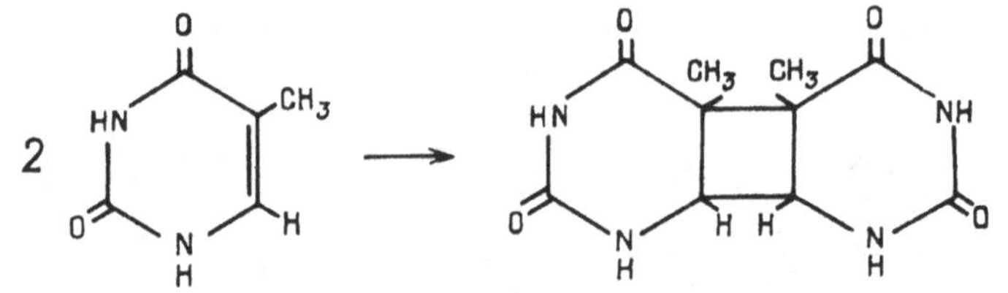

FIG. 3.31. Dimerization of thymine.

is 6–7 eV. This excitation may be transferred nonradiatively to a DNA molecule and fragment it into two strands without affecting the RNA molecule.

The two-chain DNA molecule exhibits a strong tendency to repair latent damage. Damage of single sites frequently does not disturb the function of the molecule and is rapidly repaired by duplication of the disturbed region by the neighboring undisturbed parts of the second chain. Viruses and bacteria are capable of restoring in this way large parts of the disturbed chains and to transfer a large number (10–20) of isolated damage sites.

If DNA molecules are in the cell of an organism, then enzyme repair plays an important role in their recovery in addition to the mechanisms described above. It represents a complex process involving enzymes which detect and remove damage. Effective protection to DNA from small radicals is provided by compounds containing the *S*H group. Photorepair of DNA is also possible, i.e., damage may be removed by illumination. For example, dimers are split by photons of 3–4 eV energy (see Fig. 3.32). It is assumed that an important role in this process is also played by enzymes.

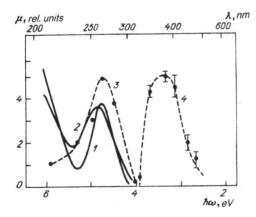

FIG. 3.32. Absorption spectra of thymine (1) and of DNA (2), and of photoformation of chromosome aberrations of Chinese hamster cells (3), and photorepair of γ-irradiated *E. coli* (4). Data taken from Refs. 42 and 128.

References

A. Basic literature

Agranovich, V. M., and Galanin, M. D., *Perenos énergii elektronnogo vozbuzhdeniya v kondensirovan- nykh sredakh* (*Transfer of Electron Excitation Energy in Condensed Media*) (Nauka, Moscow, 1978) (Sec. 2.5).*

Ansel'm, A. I., *Vvedenie v teoriyu poluprovodnikov* (Nauka, Moscow, 1978) [*Introduction to Semicon- ductor Theory* (Mir, Moscow and Prentice-Hall, Englewood Cliffs, NJ, 1981)] (Secs. 1.2, 2.2, and 2.5).

Carlson, T. A., *Photoelectron and Auger Spectroscopy* (Plenum Press, New York, 1975) (Russian trans- lation: Mir, Moscow, 1980) (Secs. 2.1, 2.2, 2.4, and 2.6).

Crawford, J. H., Jr., and Slifkin, L. M. (eds.), *Point Defects in Solids, Vol. 1, General and Ionic Crystals* (Plenum Press, New York, 1972); *Vol. 2, Semiconductors and Molecular Crystals* (Plenum Press, New York, 1975) (Russian translation edited by B. I. Boltaks, T. V. Malshovets, and A. N. Orlov, Mir, Moscow, 1979) (Secs. 3.1, 3.2, and 3.4).

Damask, D., and Dienes, G. J., *Point Defects in Metals* (Gordon and Breach, New York, 1963) (Rus- sian translation: Mir, Moscow, 1966) (Secs. 2.5 and 3.4).

Davydov, A. S., *Kvantovaya mekhanika* (Fizmatgiz, Moscow, 1963) [*Quantum Mechanics* (Pergamon Press, Oxford, 1965)] (Secs. 1.1, 1.2, 2.1, and 2.3).

Dertinger, G., and Jung, H., *Molecular Radiation Biology: Action of Ionizing Radiations on Elementary Biological Objects* (Springer Verlag, Berlin, 1970) (Russian translation: Atomizdat, Moscow, 1973) (Sec. 3.5).

Drukarev, G. F., *Stolknoveniya élektronov s atomami i molekulami* (*Collisions of Electrons with Atoms and Molecules*) (Nauka, Moscow, 1978) (Sec. 2.3).

Éidus, L. Kh., *Fiziko-kimicheskie osnovy radiobiologicheskikh protsessov i zashchita ot izluchenii* (*Phys- icochemical Basis of Radiobiological Processes and Protection from Radiation*) (Atomizdat, Moscow, 1979) (Sec. 3.5).

Fano, U., and Cooper, J. W., "Spectral distribution of atomic oscillator strengths," Rev. Mod. Phys. **40**, 441 (1968) (Sec. 2.1).

Fiermans, L., Vennik, J., and DeKeyser, W. (eds.), *Electron and Ion Spectroscopy of Solids* (Plenum Press, New York, 1978) (Russian translation: Mir, Moscow, 1981) (Secs. 1.1, 2.1, 2.2, 2.4, and 2.6).

Gantmakher, V. F., and Levinson, I. B., *Rasseyanie nositeleĭ toka v metallakh i poluprovodnikakh* (Nauka, Moscow, 1984) [*Carrier Scattering in Metals and Semiconductors* (North-Holland, Am- sterdam, 1987)] (Secs. 2.4 and 2.5).

Hasted, J. B., *Physics of Atomic Collisions* (Butterworths, London, 1964) (Russian translation: Mir, Moscow, 1965) (Secs. 2.1, 2.3, and 2.4).

Heitler, W., *The Quantum Theory of Radiation*, 3rd ed. (Clarendon Press, Oxford, 1954) (Russian translation: IL, Moscow, 1956) (Secs. 1.1, 2.1, and 2.3).

Henley, E. J., and Johnson, E. R., *Chemistry and Physics of High Energy Reactions* (Plenum Press, New York, 1969) (Russian translation: Atomizdat, Moscow, 1974) (Secs. 2.3, 3.3, and 3.5).

Kalashnikov, N. P., Remizovich, V. S., and Ryazanov, M. I., *Stolknoveniya bystrykh zaryazhennykh chastits v tverdykh telakh* (*Collisions of Fast Charged Particles in Solids*) (Atomizdat, Moscow, 1980) (Secs. 2.3, 3.1, and 3.2).

Kirsanov, V. V., Suvorov, A. L., and Trushin, Yu. V., *Protsessy radiatsionnogo defektoobrazovaniya v* metallakh (*Processes of Radiation Defect Formation in Metals*) (Énergoatomizdat, Moscow, 1985) (Secs. 3.1 and 3.2).

Migdal, A. B., and Kraĭnov, V. P., *Priblizhennye metody kvantovoĭ mekhaniki* (Nauka, Moscow, 1966) [*Approximation Methods in Quantum Mechanics*(Benjamin, New York, 1969)] (Secs. 1.1, 1.2, 2.1, and 2.3).

Mott, N. F., and Massey, H. S. W., *The Theory of Atomic Collisions*, 3rd ed. (Clarendon Press, Oxford, 1965) (Russian translation: Mir, Moscow, 1969) (Sec. 2.3).

Schiff, L. I., *Quantum Mechanics*, 2nd ed. (McGraw-Hill, New York, 1955) (Russian translation: IL, Moscow, 1959) (Secs. 1.1, 2.1, and 2.3).

Seeger, K., *Semiconductor Physics* (Springer Verlag, Berlin, 1974) (Russian translation: Mir, Moscow, 1977) (Secs. 2.2, 2.4, and 2.5).

Shul'man, A. R., and Fridrikhov, S. A., *Vtorichno-émissionnye metody issledovaniya tverdogo tela* (*Secondary-Emission Methods for Investigation of Solids*) (Nauka, Moscow, 1977) (Secs. 2.3, 2.4, and 2.6).

Thompson, M. W., *Defects and Radiation Damage in Metals* (Cambridge University Press, England, 1969) (Russian translation: Mir, Moscow, 1971) (Secs. 3.1, 3.2, and 3.4).

Vavilov, V. S., *Deĭstvie izlucheniĭ na poluprovodniki* (Fizmatgiz, Moscow, 1963) [*Effects of Radiation on Semiconductors* (Consultants Bureau, New York, 1965)] (Secs. 2.2, 2.3, 2.5, 3.1, 3.2, and 3.4).

Vavilov, V. S., Kiv, A. E., and Niyazova, O. R., *Mekhanizmy obrazovaniya i migratsii defektov v poluprovodnikakh* (*Mechanisms of Formation of Defect Migration in Semiconductors*) (Nauka, Moscow, 1981) (Secs. 3.3 and 3.4).

Vinetskiĭ, V. L., and Kholodar', G. A., *Radiatsionnaya fizika poluprovodnikov* (*Radiation Physics of Semiconductors*) (Naukova Dumka, Kiev, 1979) (Secs. 1.1, 1.3, 2.3–2.5, and 3.1–3.3).

*"(Sec. 2.5)" means that this particular reference is relevant to Sec. 2.5 of Chap. 2.

B. Cited literature

1. V. M. Akulin, V. D. Vurdov, G. G. Esadze *et al.*, Pis'ma Zh. Eksp. Teor. Fiz. **40**, 53 (1984) [JETP Lett. **40**, 783 (1984)].

2. É. D. Aluker, D. Yu. Lusis, and S. A. Chernov, *Élektronnye vozbuzhdeniya i radiolyuminestsentsiya shchelochnogalodinykh kristallov* (*Electron Excitations and Radioluminescence of Alkali Halide Crystals*) (Zinatne, Riga, 1979).

3. M. Ya. Amus'ya, M. A. Cherepkov, and L. V. Chernysheva, Zh. Eksp. Teor. Fiz. **60**, 160 (1971) [Sov. Phys. JETP **33**, 90 (1971)].

4. V. A. Andreev, V. V. Baublis, E. A. Damaskinskiĭ *et al.*, Pis'ma Zh. Eksp. Teor. Fiz. **36**, 340 (1982) [JETP Lett. **36**, 415 (1982)].

5. A. Kh. Ausmeés, Ya. Ya. Pruulmann, and M. A. Élango, Fiz. Tverd. Tela (Leningrad) **27**, 3692 (1985) [Sov. Phys. Solid State **27**, 2225 (1985)].

6. V. G. Baru and F. F. Vol'kenshteĭn, *Vliyanie oblucheniya na poverkhnostnye svoĭstva poluprovodnikov* (*Influence of Irradiation on Surface Properties of Semiconductors*) (Nauka, Moscow, 1978).

7. N. G. Basov, O. V. Bogdankevich, P. G. Eliseev, and B. M. Lavrushkin, Fiz. Tverd. Tela (Leningrad) **8**, 1341 (1966) [Sov. Phys. Solid State **8**, 1073 (1966)].

8. N. G. Basov, O. V. Bogdankevich, and B. M. Lavrushkin, Fiz. Tverd. Tela (Leningrad) **8**, 21 (1966) [Sov. Phys. Solid State **8**, 15 (1966)].

9. L. Z. Benimetskaya, A. L. Kozionov, S. Yu. Novozhilov, and M. I. Shtokman, Dokl. Akad. Nauk SSSR **272**, 217 (1983).

10. V. A. Boĭko, F. V. Bunkin, V. I. Derzhiev, and S. I. Yakovlenko, Izv. Akad. Nauk SSSR Ser. Fiz. **47**, 1880 (1983).

11. V. N. Brudnyĭ, A. A. Groza, and M. A. Krivov, Izv. Vyssh. Uchebn. Zaved. Fiz. No. 4, 101 (1982).

12. V. S. Vavilov, L. K. Vodop'yanov, and N. I. Kurdiani, *Radiatsionnaya fizika nemetallicheskikh kristallov* (*Radiation Physics of Nonmetallic Crystals*) (Naukova Dumka, Kiev, 1967), p. 191.

13. D. I. Vaĭsburd, B. N. Semin, É. G. Tavanov *et al.*, *Vysokoénergeticheskaya élektronika tverdogo tela* (*High-Energy Electronics of Solids*) (Nauka, Novosibirsk, 1982).

14. V. L. Vinetskiĭ and G. A. Kholodar', *Statisticheskoe vzaimodeĭstvie élektronov i defektov v poluprovodnikakh (Statistical Interaction of Electrons and Defects in Semiconductors)* (Naukova Dumka, Kiev, 1969).

15. V. O. Vyazemskiĭ, I. I. Lomonosov, A. N. Pisarevskiĭ *et al.*, *Stsintilyatsionnyĭ metod v radiometrii (Scintillation Method in Radiometry)* (Atomizdat, Moscow, 1961).

16. A. A. Gaĭlitis, Uch. Zap. Latv. Gos. Univ. **234** (3), 26 (1975).

17. V. V. Gan and O. V. Yudin, Vopr. At. Nauk Tekh. Ser. Fiz. Radiats. Povrezhd. Radiats. Materialoved. No. 2 (25), 11 (1983).

18. A. G. Gaydon, *Dissociation Energies and Spectra of Diatomic Molecules* (Chapman and Hall, London, 1947) (Russian translation: IL, Moscow, 1949).

19. V. I. Gol'danskiĭ and Yu. M. Kagan, Usp. Fiz. Nauk **110**, 445 (1973) [Sov. Phys. Usp. **16**, 563 (1974)].

20. A. M. Gurvich, *Rentgenolyuminofory i rentgenovskie ékrany (X-Ray Phosphors and X-Ray Screens)* (Atomizdat, Moscow, 1976).

21. V. P. Denks, A. É. Dudel'zak, Ch. B. Lushchik *et al.*, Zh. Prikl. Spektrosk. **24**, 37 (1976).

22. Yu. N. Dmitriev, V. N. Kulik, L. P. Gal'chinetskiĭ, and V. M. Koshkin, Fiz. Tverd. Tela (Leningrad) **17**, 3685 (1975) [Sov. Phys. Solid State **17**, 2396 (1975)].

23. V. V. Emtsev, M. I. Klinger, T. V. Mashovets *et al.*, Fiz. Tekh. Poluprovodn. **13**, 933 (1979) [Sov. Phys. Semicond. **13**, 546 (1979)].

24. V. I. Zemskiĭ, B. P. Zakharchenya, and D. N. Mirlin, Pis'ma Zh. Eksp. Teor. Fiz. **24**, 96 (1976) [JETP Lett. **24**, 82 (1976)].

25. É. R. Il'mas, G. G. Liĭd'ya, and Ch. B. Lushchik, Opt. Spektrosk. **18**, 453 (1965) [Opt. Spectrosc. (USSR) **18**, 255 (1965)].

26. Yu. Kagan, L. A. Maksimov, and N. V. Prokof'ev, Pis'ma Zh. Eksp. Teor. Fiz. **36**, 204 (1982) [JETP Lett. **36**, 253 (1982)].

27. Yu. Kh. Kalnin' and E. A. Kotomin, Vopr. At. Nauk Tekh. Ser. Fiz. Radiats. Povrezhd. Radiats. Materialoved. No. 1 (29), 18 (1984).

28. N. V. Karlov and A. M. Prokhorov, Usp. Fiz. Nauk **118**, 583 (1976) [Sov. Phys. Usp. **19**, 285 (1976)].

29. T. A. Carlson, *Photoelectron and Auger Spectroscopy* (Plenum Press, New York, 1975) (Russian translation: Mir, Moscow, 1980).

30. M. I. Klinger, Ch. B. Lushchik, T. V. Mashovets, G. A. Kholodar', M. K. Sheĭnkman, and M. A. Elango, Usp. Fiz. Nauk **147**, 523 (1985) [Sov. Phys. Usp. **28**, 994 (1985)].

31. M. P. Klyap, V. A. Kritskiĭ, Yu. A. Kulyupin *et al.*, Zh. Eksp. Teor. Fiz. **86**, 1117 (1984) [Sov. Phys. JETP **59**, 653 (1984)].

32. I. S. Kolomnikov, T. V. Lysyak, E. A. Konash *et al.*, Dokl. Akad. Nauk SSSR **265**, 912 (1982).

33. E. A. Konorova, S. F. Kozlov, and V. S. Vavilov, Fiz. Tverd. Tela (Leningrad) **8**, 3 (1966) [Sov. Phys. Solid State **8**, 1 (1966)].

34. V. Yu. Karasov, S. N. Shamin, V. A. Nikolaenko *et al.*, Fiz. Tverd. Tela (Leningrad) **26**, 2873 (1984) [Sov. Phys. Solid State **26**, 1739 (1984)].

35. I. L. Kuusmann, P. Kh. Liblik, and Ch. B. Lushchik, Pis'ma Zh. Eksp. Teor. Fiz. **21**, 161 (1975) [JETP Lett. **21**, 72 (1975)].

36. O. I. Leĭpunskiĭ, V. V. Novozhilov, and V. N. Sakharov, *Rasprostranenie gamma-kvantov v veshchestve (Propagation of Gamma Photons in Matter)* (Fizmatgiz, Moscow, 1960).

37. V. S. Letokhov, *Nelineĭnye selektivnye fotoprotsessy v atomakh i molekulakh (Nonlinear Selective Photoprocesses in Atoms and Molecules)* (Nauka, Moscow, 1983).

38. Ch. B. Lushchik, I. K. Vitol, and M. A. Élango, Usp. Fiz. Nauk **122**, 223 (1977) [Sov. Phys. Usp. **20**, 489 (1977)].

39. A. A. Maĭste, R. É. Ruus, and M. A. Élango, Zh. Eksp. Teor. Fiz. **79**, 1671 (1980) [Sov. Phys. JETP **52**, 844 (1980)].

40. A. A. Manenkov and A. M. Prokhorov, Usp. Fiz. Nauk **148**, 179 (1986) [Sov. Phys. Usp. **29**, 104 (1986)].

41. V. V. Nemoshkolenko and V. G. Aleshin, *Élektronnaya spektroskopiya kristallov (Electron Spectroscopy of Crystals)* (Naukova Dumka, Kiev, 1976).

42. D. N. Nikogosyan, Priroda No. 2, 77 (1982).

43. É. L. Nolle, V. S. Vavilov, G. P. Golubev, and V. S. Mashtakov, Fiz. Tverd. Tela (Leningrad) **8**, 286 (1966) [Sov. Phys. Solid State **8**, 236 (1966)].
44. I. A. Parfianovich and É. É. Penzina, *Élektronnye tsentry okraski v ionnykh kristallakh (Electron Color Centers in Ionic Crystals)* (East-Siberian Book Press, Irkutsk, 1977).
45. E. R. Peressini, in: *Laser Radiation Effects* (Russian translation: Mir, Moscow, 1968), p. 132.
46. A. K. Pikaev, B. G. Ershov, and O. Kh. Khodzhaev, in: *Proceedings of Interuniversity Conference on Radiation Physics* (Tomsk State University, 1970), p. 333.
47. V. G. Plekhanov and A. A. O'Connel-Bronin, Pis'ma Zh. Eksp. Teor. Fiz. **27**, 30 (1978) [JETP Lett. **27**, 27 (1978)].
48. K. K. Rebane, Tr. Inst. Fiz. Astron. Akad. Nauk Est. SSR **7**, 62 (1958).
49. K. K. Rebane, Zh. Prikl. Spektrosk. **37**, 906 (1982).
50. I. Yu. Tekhver and V. V. Khizhnyakov, Zh. Eksp. Teor. Fiz. **69**, 599 (1975) [Sov. Phys. JETP **42**, 305 (1975)].
51. M. W. Thompson, *Defects and Radiation Damage in Metals* (Cambridge University Press, 1969) (Russian translation: Mir, Moscow, 1971).
52. J. P. Phillips, *Spectra-Structure Correlations* (Academic Press, New York, 1964) (Russian translation: Mir, Moscow, 1968).
53. N. A. Tsal', Yu. V. Karavan, R. I. Didyk, and O. P. Dragan, Dokl. Akad. Nauk SSSR **220**, 658 (1975).
54. A. M. Shalaev and A. A. Adamenko, *Radiatsionno-stimulirovannoe izmenenie élektronno struktury metallov (Radiation-Stimulated Changes in the Electron Structure of Metals)* (Atomizdat, Moscow, 1977).
55. K. K. Shvarts, Z. A. Grant, M. M. Grube, and T. K. Mezhs, *Termolyuminestsentnaya dozimetriya (Thermoluminescence Dosimetry)* (Zinatne, Riga, 1968).
56. K. K. Shvarts and Yu. A. Ékmanis, Radiats. Fiz. **4**, 11 (1966).
57. L. Kh. Éidus, *Fiziko-khimicheskie osnovy radiobiologicheskikh protsessov i zashchita ot izluchenii (Physicochemical Basis of Radiobiological Processes and Protection from Radiation)* (Atomizdat, Moscow, 1979) (Sec. 3.5).
58. M. A. Élango, Fiz. Tverd. Tela (Leningrad) **17**, 2356 (1975) [Sov. Phys. Solid State **17**, 1555 (1975)].
59. A. Akilbekov, A. Dauletbekova, and A. Elango, Phys. Status Solidi B **127**, 493 (1985).
60. R. C. Alig and S. Bloom, Phys. Rev. Lett. **35**, 1522 (1975).
61. A. Ausmees, M. Elango, A. Kikas, and J. Pruulmann, Phys. Status Solidi B **137**, 495 (1986).
62. J. L. Bahr, Contemp. Phys. **14**, 329 (1973).
63. R. Balzer, H. Peisl, and W. Waidelich, in: *Proceedings of International Symposium on Color Centers in Alkali Halides*, Rome, 1968, p. 20.
64. W. Bambynek, B. Crasemann, R. W. Fink *et al.*, Rev. Mod. Phys. **44**, 716 (1972).
65. G. Baur, F. Rösel, and D. Trautmann, J. Phys. B **16**, L419 (1983).
66. H. B. Bebb and A. Gold, Phys. Rev. **143**, 1 (1966).
67. E. Bogh and J. L. Whitton, Phys. Rev. Lett. **19**, 553 (1967).
68. W. L. Brown and W. M. Augustyniak, J. Appl. Phys. **30**, 1300 (1959).
69. T. A. Carlson, W. E. Hunt, and M. O. Krause, Phys. Rev. **151**, 41 (1966).
70. T. A. Carlson and M. O. Krause, Phys. Rev. **137**, A1655 (1965).
71. T. A. Carlson and R. M. White, J. Chem. Phys. **44**, 4510 (1966).
72. K. Codling, J. Electron. Spectrosc. Relat. Phenom. **17**, 279 (1979).
73. H. G. Cooper, J. S. Koehler, and J. W. Marx, Phys. Rev. **97**, 599 (1955).
74. J. W. Corbett, Jr. and J. C. Bourgoin, in: *Point Defects in Solids, Vol. 2, Semiconductors and Molecular Crystals*, edited by J. H. Crawford, Jr. and L. M. Slifkin (Plenum Press, New York, 1975), p. 1.
75. S. Datz, C. D. Moak, O. H. Crawford *et al.*, Phys. Rev. Lett. **40**, 843 (1978).
76. M. A. Duguay, in: *Laser Induced Fusion and X-Ray Laser Studies* (Proceedings of Conference, Santa Fe, NM, 1975, edited by S. F. Jacobs, M. O. Scully, and M. Sargent), publ. as *Physics of Quantum Electronics*, Vol. 3 (Addison-Wesley, Reading, MA, 1976), p. 557.
77. W. Eberhardt, T. K. Sham, R. Carr *et al.*, Phys. Rev. Lett. **50**, 1038 (1983).
78. A. A. Elango and T. N. Nurakhmetov, Phys. Status Solidi B **78**, 529 (1976).

79. M. Elango, A. Kikas, E. Nõmmiste, J. Pruulmann, and A. Saar, Phys. Status Solidi B **114**, 487 (1982).
80. M. Elango and A. E. Kiv, Cryst. Lattice Defects Amorphous Mater. **11**, 305 (1986).
81. M. Elango, J. Pruulmann, and A. P. Zhurakovski, Phys. Status Solidi B **115**, 399 (1983).
82. J. H. Evans, Nature (London) **229**, 403 (1971).
83. L. Federici, G. Giordano, G. Matone *et al.*, Nuovo Cimento B **59**, 247 (1980).
84. H. R. Fetterman, J. A. Davis, and D. B. Fitchen, in: *Proceedings of International Symposium on Color Centers in Alkali Halides*, Rome, 1968, p. 93.
85. R. L. Fleischer, P. B. Price, and R. M. Walker, J. Appl. Phys. **36**, 3645 (1965).
86. F. Gerkern, J. Barth, K. L. I. Kobayashi, and C. Kunz, Solid State Commun. **35**, 179 (1980).
87. J. B. Gibson, A. N. Goland, M. Milgram, and G. H. Vineyard, Phys. Rev. **120**, 1229 (1960).
88. A. A. Gorokhovski and J. Kikas, Opt. Commun. **21**, 272 (1977).
89. R. Haensel, G. Keitel, P. Schreiber, and C. Kunz, Phys. Rev. **188**, 1375 (1969).
90. N. Hayashi and T. Takahashi, Appl. Phys. Lett. **41**, 1100 (1982).
91. F. J. de Heer, R. H. J. Jansen, and W. van der Kaay, J. Phys. B **12**, 979 (1979).
92. F. Magnotta, I. P. Herman, and F. I. Aldridge, Chem. Phys. Lett. **92**, 600 (1982).
93. D. R. Hertling, R. K. Feeney, D. W. Hughes, and W. E. Sayle, J. Appl. Phys. **53**, 5427 (1982).
94. R. C. Hughes, Phys. Rev. B **15**, 2012 (1977).
95. A. Hustrulid, P. Kusch, and J. T. Tate, Phys. Rev. **54**, 1037 (1938).
96. N. Itoh, Cryst. Lattice Defects **3**, 115 (1972).
97. J. B. Johnson and K. G. McKay, Phys. Rev. **91**, 582 (1953).
98. H. Käämbre and V. Bichevin, in: *Abstracts of Papers presented at International Conference on Color Centres in Ionic Crystals*, Reading, England, 1971, Abstr. No. 146.
99. M. N. Kabler and R. T. Williams, Phys. Rev. B **18**, 1948 (1978).
100. J. Kikas and M. Elango, Phys. Status Solidi B **130**, 211 (1985).
101. R. K. Klein, J. O. Kephart, R. H. Pantell *et al.*, Phys. Rev. B **31**, 68 (1985).
102. M. I. Klinger and T. V. Mashovets, Cryst. Lattice Defects **9**, 113 (1981).
103. A. R. Knudson, P. G. Burkhalter, and D. J. Nagel, in: *Proceedings of International Conference on Inner Shell Ionization Phenomena and Future Applications*, Oak Ridge, TN, 1972, Vol. 3 (1973), p. 1675.
104. S. E. Kohn, P. Y. Yu, Y. Petroff *et al.*, Phys. Rev. B **8**, 1477 (1973).
105. Ya. S. Lebedev, Radiat. Eff. **1**, 213 (1969).
106. V. Lotz, Z. Phys. **206**, 205 (1967).
107. G. V. Marr and J. B. West, At. Data Nucl. Data Tables **18**, 497 (1976).
108. O. J. Marsh, R. Baron, G. A. Shifrin, and J. W. Mayer, Appl. Phys. Lett. **13**, 199 (1968).
109. D. L. Matthews, P. L. Hagelstein, M. D. Rosen *et al.*, Phys. Rev. Lett. **54**, 110 (1985).
110. W. C. McGowan and W. A. Sibley, Philos. Mag. **19**, 967 (1969).
111. W. A. Metz and E. W. Thomas, J. Appl. Phys. **53**, 3529 (1982).
112. J. D. Meyer and B. Stritzker, Phys. Rev. Lett. **48**, 502 (1982).
113. H. Morita, A. Chamberod, and S. Steinemann, J. Phys. F **14**, 3053 (1984).
114. P. W. Palmberg and T. N. Rhodin, J. Phys. Chem. Solids **29**, 1917 (1968).
115. C. C. Parks, Z. Hussain, D. A. Shirley *et al.*, Phys. Rev. B **28**, 4793 (1983).
116. V. N. Pavlovich, Phys. Status Solidi B **116**, K9 (1983).
117. M. F. Perutz, S. S. Hasnain, P. J. Duke *et al.*, Preprint No. DL/SCI/P302E (Daresbury Nuclear Laboratory, England, 1981), p. 1.
118. W. C. Price and A. W. Potts, in: *Proceedings of International Symposium for Synchrotron Radiation Usage*, Daresbury, 1973 (Daresbury Nuclear Laboratory, England, 1973), p. 190.
119. K. Radler and B. Sonntag, Chem. Phys. Lett. **39**, 371 (1976).
120. K. V. Reddy and M. J. Berry, Chem. Phys. Lett. **66**, 223 (1979).
121. R. H. Ritchie, C. J. Tung, V. E. Anderson, and J. C. Ashley, Radiat. Res. **64**, 181 (1975).
122. M. T. Robinson and I. M. Torrens, Phys. Rev. B **9**, 5008 (1974).
123. K. Sattler, J. Mühlbach, O. Echt *et al.*, Phys. Rev. Lett. **47**, 160 (1981).
124. M. P. Seah, Surf. Interface Anal. **1**, 86 (1979).
125. K. Siegbahn, K. Nordling, A. Fahlman *et al.*, *ESCA—Atomic, Molecular, and Solid State Structure Studied by Means of Electron Spectroscopy* (Almquist and Wiksells, Uppsala, Sweden, 1967).
126. P. J. Siemens and J. O. Rasmussen, Phys. Rev. Lett. **42**, 880 (1979).

127. E. A. Silinsh and A. J. Jurgis, Chem. Phys. **94**, 77 (1985).

128. V. G. Skvortsov, M. N. Myasnick, V. A. Sokolov, and I. I. Morozov, Photochem. Photobiol. **33**, 187 (1981).

129. B. M. J. Smets and T. P. A. Lommen, J. Am. Ceram. Soc. **65**, 80 (1982).

130. K. Sonnenberg, W. Schilling, K. Mika, and K. Dettmann, Radiat. Eff. **16**, 65 (1972).

131. M. Sparks, D. L. Mills, R. Warren *et al.*, Phys. Rev. B **24**, 3519 (1981).

132. H. Suzuki, K. Ishikawa, and T. Yoshihara, J. Phys. Soc. Jpn. **51**, 3643 (1982).

133. J. H. O. Varley, J. Nucl. Energy **1**, 130 (1954).

134. N. R. Whetten and A. B. Laponsky, Phys. Rev. **107**, 1521 (1957).

135. C. W. White, W. H. Christie, B. R. Appleton *et al.*, Appl. Phys. Lett. **33**, 662 (1978).

136. M. Yoshida, J. Phys. Soc. Jpn. **16**, 44 (1961).

Some physical constants

Velocity of light	c	3.00×10^8 m/s
Planck constant	h	6.63×10^{-34} J s
Planck–Dirac constant	$\hbar$	1.05×10^{-34} J s
Electron charge	e	1.60×10^{-19} C
Electron rest mass	m	9.11×10^{-31} kg
Proton and neutron rest masses	M_p	1.67×10^{-27} kg
Avogadro number	N_A	6.02×10^{23} mol^{-1}
Boltzmann constant	k_B	1.38×10^{-23} J/K
Fine-structure constant	$\alpha = \dfrac{e^2}{\hbar c}$	7.30×10^{-3}
	α^{-1}	137.0
Thomson scattering cross section	$8/3 \pi r_0^2$	6.57×10^{-29} m^2
Atomic length (Bohr radius)	$a_B = \dfrac{\hbar^2}{me^2}$	5.29×10^{-11} m
Atomic velocity	$v_B = \dfrac{e^2}{\hbar} = \alpha c$	2.19×10^6 m/s
Atomic time	$\tau_B = \dfrac{a_B}{v_B}$	2.42×10^{-17} s
Atomic frequency	$\omega_B = \tau_B^{-1}$	4.13×10^{16} s^{-1}
Atomic energy	$E_B = \dfrac{e^2}{a_B} = \dfrac{me^4}{\hbar^2}$	4.36×10^{-18} J
Classical electron radius	$r_0 = \dfrac{e^2}{mc^2} = \alpha^2 a_B$	2.82×10^{-15} m
Electron Compton wavelength	$\lambda_c = \dfrac{\hbar}{mc}$	2.43×10^{-12} m
Electron rest energy	mc^2	8.18×10^{-14} J
Unit of energy	1 eV	1.60×10^{-19} J

Index

143

Printed in the United States
By Bookmasters